# Cross-Cultural Neuropsychological Assessment: Theory and Practice

# Cross-Cultural Neuropsychological Assessment: Theory and Practice

Victor Nell
*University of South Africa*

LEA LAWRENCE ERLBAUM ASSOCIATES, PUBLISHERS
2000 Mahwah, New Jersey London

Lawrence Erlbaum Associates, Inc., Publishers
10 Industrial Avenue
Mahwah, New Jersey 07430

**Library of Congress Cataloging-in-Publication Data**

Nell, V.
Cross-cultural neuropsychological assessment : theory and practice / Victor Nell.
p. cm.
Includes bibliographical references and index.
ISBN 0-8058-3355-2 (cloth : alk. paper). — ISBN 0-8058-3356-0 (pbk. : alk. paper)
1. Neurologic examination. 2. Transcultural medical care.
I. Title.
RC348.N45 1999
616.8'0475—dc21 99-32682
CIP

Books published by Lawrence Erlbaum Associates are printed on acid-free paper, and their bindings are chosen for strength and durability.

Printed in the United States of America
10 9 8 7 6 5 4 3 2 1

*To Myrna,*
*for the second time*

# Contents

## PART III THE PRACTICE OF CROSS-CULTURAL ASSESSMENT

## FIGURES

## TABLES

## APPENDICES

# Preface

This book is for all neuropsychologists who are called upon to assess culturally different clients: With very few exceptions, this means every neuropsychologist. Throughout the Western world there are migrant and refugee minorities whose home language and culture differ from those of the host country. Many suburbs and workplaces in Western countries, far from being culturally and linguistically homogenous, are more like a meeting of the United Nations General Assembly. While I was writing this preface, a colleague in Minnesota told me that over the past few years he has assessed clients from Egypt, Algeria, Liberia, Iraq, and Iran, as well as Hmong people from Laos and elderly Russian Jews. A few years ago, he continued, this American heartland was so culturally homogenous that cross-cultural assessment was not an issue.

Today, not only in Minnesota but throughout the Western world, migration raises assessment and test validity problems that cannot be ignored. In the southern hemisphere, these problems are pervasive and demand a solution even more imperatively. But it is not only migration that makes cross-cultural skills imperative.

In the West, it seems quite obvious that the problems arising from cultural difference do not relate to native-born ethnic minorities whose first language is the majority language and who are educated in the mainstream schooling system. Thus, at first glance, America's Irish, Jews, and African Americans are Western. But, at second glance, what of the tight-knit community of Yiddish-speaking Hassidic Jews in Crown Heights, NY? And, why do some apparently mainstream groups in the United States score significantly lower than the mean on standard IQ tests?

When it comes to culture, nothing is easy or clear-cut. If there is one big lesson to be learned from the IQ controversy (chap. 4), it is that until language proficiency, quality of education, test-wiseness, cognitive style, and the other components of acculturation have been proved beyond any reasonable doubt to be equivalent for the groups whose scores are being compared, score differences cannot be attributed to genetic differences.

* * *

Of course, there would be no need for this book if there were good test norms for all populations. Unfortunately, there are not, and it is unlikely that there will be in the foreseeable future. There are no "norms" until construct validity for the population in question has been demonstrated. Demonstrating construct validity in turn requires a large body of scores for a representative sample of that population (chap. 5). Providing norms is therefore a slow and costly exercise. It is not commercially viable unless there is a large homogenous target population, as for example in North America and Europe. But for the fragmented cultural enclaves and myriad languages of Africa, South America, and other countries, test standardization will not be the product of market forces, which are rapid and efficient, but rather the result of a slow growth of local research initiatives.

What follows is an attempt to deal realistically with the problem of doing neuropsychological assessments without reliable norms, or with no norms at all.

* * *

If the purpose is so simple and practical, then why is the whole of Part II—four dense and difficult chapters—given over to theory? Clinicians faced with pressing real-world assessment problems will be inclined to skip this section and go straight on to Part III. However, neuropsychological assessment is an abstract and complex activity that is acutely sensitive, consciously and unconsciously, to its theoretical underpinnings. The historical, exotic, and technical material in these chapters is intensely practical, with the potential to change the way clinicians interpret their assessment findings and structure their reports. I believe the complexity is necessary, and hope that readers will in the end come to agree with me.

So much for what this book is: a neuropsychological assessment method that accommodates to cultural difference and the lack of appropriate norms. It is not a standard reference or primer. It assumes that readers are familiar with the manifestations of neuropsychological deficits and with neuropsychological tests; its method is based on qualitative interpretation, but it does not seek to recapitulate the content of the excellent existing texts on these methods.

And it is not a last word. Neuropsychologists who have assessed people from other cultures know that a fundamentally new conceptualization of the assessment process is needed. The material in chapters 9 and 10 responds to that need by setting out what I believe to be the best assessment method that can be derived from current theory. The method is exploratory and needs to be validated, but the interests of culturally different clients will be better served by setting this process in motion on as wide a front as possible: Controversy and the progress controversy brings are preferable to continued stagnation.

## SYNOPSIS

Parts I and II of the book set out the need for the assessment method described in Part III.

Part I frames the problem. Chapter 1 specifies the antecedents and consequences of test-wiseness, and then relates testing to the politics of culture, race, and class by showing how psychology, for many years, succumbed to racist beliefs. Chapter 2 argues that neuropsychology's claim to universalism derives from incorrect assumptions about the performance equivalence of all people everywhere, and shows that test score differences between countries and cultures leave the universalist claim threadbare.

Part II sets out the theoretical principles on which cross-cultural assessment is based. It asks why scores differ so markedly across cultures, and examines some of the competing answers to the question: environmentalism (chap. 3) on the one hand, and the hereditarian view on the other, as well as the political issues embedded within the IQ controversy (chap. 4). Chapter 5 unpacks the concept of construct validity and sets out a method of establishing cross-cultural construct equivalence. Chapter 6 argues for a behavioral assessment method that is latent in all neuropsychology; in underresourced settings, this behavioral neuropsychology is a readily transferable technology that opens the way to the wider dissemination of neuropsychological diagnosis and treatment.

Part III is practical. Chapters 7 and 8 set out the content and method of the behavioral assessment; chapters 9 and 10 do the same for the testing process. Chapter 9 builds on Vygotsky's notion of the zone of proximal development in order to set in motion a testing method that is kindly and helpful rather than intimidating and distant. Truly matching the performance of culturally different clients (for whom testing is scary and test content alien) to their potential is a distant goal—but the "buds, flowers, fruits" approach set out in this chapter is a helpful step in that direction. Chapter 10 argues for a core test battery consisting of a small group of often-used tests as a necessary step toward the goal of construct determination, and reviews the tests that have been chosen in relation to the instructional and practice needs of non-test-wise clients. The final chapter shows the utility of organizing reports around a behavioral frame within which the test results are accommodated.

## ACKNOWLEDGMENTS

This book has been a long time in the making. In 1978, I borrowed a copy of the first edition of Muriel Lezak's *Neuropsychological Assessment* from my PhD supervisor, Dev Griesel, and discovered that a wonderful new world lay out there beyond the clicks and flashes of the psychophysiological laboratory. The very word *neuropsychology* had a magical quality. A year later, I apprenticed myself to Harry Reef, professor of neurology at the Johannesburg Hospital, who introduced me to the first cases I had ever seen, including a man with dressing apraxia whose face—and bewilderment—I can still see before me; my first case report, applying Christensen's version of Luria's neuropsychological examination to a client with an expressive aphasia, ran to 47 pages. Harry blanched when I handed it to him. Two years later, I did a postdoctoral fellowship under the benign supervision of Chuck Matthews in Madison, Wisconsin, where I learned diagnostics, coherence, and brevity. Over the years, especially during his term as president of the International Neuropsychological Society, Chuck gave influential support to the view that clinical neuropsychology needed a wider base than the North American.

Since then, I have incurred many other debts. In South Africa, one works in exile, but among friends. I am grateful first to my colleagues in neuropsychology in the Health Psychology Unit (Jacqui Sesel, Digby Brown, and Johan Kruger), and to Gail Barton, who with amazing good humor produced many drafts. Terry Taylor gave shape to the notion of bypassing culture. The collaboration of Jonny Myers and Leslie London at the University of Cape Town generated most of the South African data reported in chapter 2. In the text, "our group" refers to the collaboration between these individuals and institutions. Thanks also to Stuart Anderson, who made prepublication scores available to me, and to several generations of students for other scores, especially Nomahlubi Makunga, Ndepu Moselenyane, and Mandla Adonisi. I thank the University of South Africa and my department for generous leave, and their comforting faith that the book would in the fullness of time be completed. I also thank James Kitching, who served as a gateway to our wonderful university library. The little group that founded the South African Clinical Neuropsychology Association also nurtured my own commitment to neuropsychology in South Africa: Rory Plunkett, Shirley Tollman, and (back then) Mike Saling. I thank the American neuropsychologists who asked searching questions about the book's identity and assessment method: Ken Adams, Steve Sarfaty, Dennis Johnston, Harold Maphet, and Tom Kay. I learned amazing things about psychotherapy with brain-injured clients from Esti Klag, Alex Butchart, and Brian Mallinson, from the clients[1] themselves at our unit's Outpatient Concussion Clinic at Hillbrow Hospital in Johannesburg, and before that at the Edenvale and Sterkfontein Hospitals. Long-suffering or rightfully cantankerous, they endured my errors and arrogance, now tempered by uncertainty. Thanks also to Roy Sugarman, once a student and now a friend; to my children and their friends, willing and unwilling test subjects; to Myrna, whose book this also is; and to our dogs, who spent one glorious sabbatical on the Kommetjie seashore in Cape Town while the first draft was written, and another patiently waiting for us to reappear from Princeton, New Jersey, where the book was completed.

Victor Nell
email: nellv@iafrica.com

[1]I have throughout used *client*, unless referring to research studies or citing the work of others, when "subject" is used.

Part I

# THE SEDUCTION OF UNIVERSALS

Chapter 1

# Westernization and the Politics of Culture

Psychological tests are conceived and standardized within the matrix of Western culture. This book focuses on individuals whose culture is not the same as the culture of the test-maker or of the test-maker's target group, which is almost invariably Western. Geographically, "Western" refers to the countries of western Europe and North America. This seems strange: What cultural commonality might one attribute to London, Stockholm, New York, Rome, and Quebec? Although they do not have a common language, they do share the cultural tradition that Bloom (1994) called the Western canon—a body of literature that includes works in English, French, Italian, Spanish, German, and Russian, and is authoritative in Western culture.

There are also psychological commonalities that derive from the ascendancy of individualism over collectivism (Triandis, 1995), and the centrality of the achievement motive (McClelland, 1961) in Western culture. Westerners have a driven, competitive attitude not only to testing, described later, but also to all other opportunities for the demonstration of individual excellence. Brooding over these values is the Protestant Ethic (Albee, 1977; Weber, 1904/1965), a secular religion that brings salvation in the next world and high status in this one through hard work and unremitting effort.

But the core psychological meaning of Westernization is "test-wiseness." Test-taking skills are so taken for granted in Western society that it is difficult to grasp the extent to which they are absorbed rather than explicitly taught. We all grow up knowing that when you take a test, you are highly motivated, that is, keyed up, a little nervous, and ready (with not a little trepidation) to meet the challenge. Consequently, when the test session begins, you sit still, concentrate intensely, don't chat with the examiner (even in a one-to-one situation), and take it for granted even without being told—though the tester will nonetheless often tell you—that you must work as fast and accurately as you can.

How absurd! "Accurate" means slow and careful; "fast" means that quantity counts more than quality. How does one bring these contradictory demands together in a single smooth performance? Yet we do—by dint of intense effort, and rapid scanning back for errors even as the next item is begun, with the help of Western society's treasured article of faith: that smart is fast (Robert Sternberg's phrase) and that in the trade-off between speed and power, speed must win. For bearers of other cultures, in which smart on the contrary means cautious and thoughtful, this instruction will likely be dismissed for the nonsense it is, and customary cultural attitudes will be brought to bear on the novel test situation.

Test-wiseness is most powerfully acquired through the formal educational system. In the developed countries, a high school education (10 or 12 years of formal schooling) is sufficient to guarantee a well-practiced repertoire of test-taking skills. These skills are imparted both directly by multiple experiences with IQ, aptitude, and streaming tests, and the obvious similarities between these and school examinations, and indirectly, because success on psychological tests draws on behaviors valued in the classroom (i.e., sitting still, paying attention for long periods, using pen and pencil dextrously, copying designs and solving problems, and working fast to keep up with the class and to finish examinations on time). In the developing countries, where high schools are underfunded, poorly equipped, and without the resources for psychological testing or hands-on laboratory work (Rogan & MacDonald, 1983), teachers may not themselves be bearers of Western culture, and with a high pupil–teacher ratio, a high school education may not be enough to entrench test-wiseness.

Matters are further complicated by clients' socioeconomic status—that is, the relative cultural richness or deprivation of their background, which is in turn greatly influenced by urban or rural location. Indeed, as noted in chapter 10, urbanization and schooling are the two major conduits for the absorption of the Western attitudes to test-taking already described.

## LITERACY AS A SUMMATIVE CRITERION

At this array of intersections between Western culture and clients who are to a greater or lesser extent external to that culture, how can these judgment calls between "Western" and "culturally different" be made? The gold standard is that the constructs underlying tests—and, for that matter, interview questions—must have a common, shared existence in the minds of the test-maker and the test-taker. With Western clients—loose as that term is—questions about the construct validity of tasks such as clock drawing, Raven's Progressive Matrices, or Trails B do not arise—not in Rome or Quebec or Stockholm. It is taken for granted that the constructs exist equally in both the minds in question, and there is no evidence to the contrary—not I hasten to add that such evidence has been energetically sought: Matters might be more complicated than we would like to think.

With clients from other cultures—and this criterion has nothing to do with skin color or country of residence—the absence of construct equivalence is sometimes clear, as with Luria's Uzbekhis in the 1930s (chap. 3) and with subsistence farmers in the developing countries today. More often, the examiner will be uncertain about a client's "cultural location."

In such cases, as a summative criterion, literacy level is for the time being the best and most readily accessible indicator of Westernization. Persons who are not fully literate are variously described as illiterate, functionally illiterate, or semiliterate (United Nations Educational, Scientific, and Cultural Organization [UNESCO], 1957). *Illiteracy* among urbanized people is rare, because nearly everyone can identify numbers and name them fluently, and also letters, although with less fluency. *Functional illiteracy* refers to individuals who can sign their name and read a text at grade school level, but who for example are unable independently to follow a recipe or the instruction sheet enclosed with a product. The term *semiliterate* is used both for functionally illiterate individuals and those who exceed this definition in that they can say the alphabet, albeit nonfluently, and comprehend simple written materials. However, they cannot fluently perform tasks like generating words to a given letter, serial subtraction, or the analogical reasoning required for multistep arithmetic problems and syllogisms—all tasks that are fully developed in high school.

In summary (these matters are more fully explored in chaps. 4 and 10), the threshold I recommend for achievement of functional literacy is completion of high school (i.e., 12 years of formal education). This minimum is required for tests requiring well-entrenched numeric, reading, or analogical reasoning skills, such as the Paced Auditory Serial Addition Test (PASAT), the Stroop, or the Wisconsin Card Sorting Test (WCST). This is a more stringent standard than usual, because all Western adults, even those without a high school education, are scored on the Wechsler Adult Intelligence Scale–Revised (WAIS–R); however, in the West, this lack of formal schooling is partially compensated for by the intense socialization and acculturation experiences offered by technological, media-rich environments.

The clients for whom the methods set out in Part III are intended are thus semiliterate persons from culturally different environments, and also native-born Westerners who are not fully test-wise, or for whom the construct equivalence of tests is in doubt. For all such clients, there are likely to be substantial benefits if the guided learning assessment method described in chapter 9 is applied.

Of course, the peoples of Asia, North Africa, and the Middle East are as culturally different—or non-Western—as those of Africa and South America. Yet, this book is much more about sub-Saharan Africa than it is a truly global book. Given this limitation, is this book relevant for all culturally different clients? The answer is "no" in that the clinical content of Part III is far from universal. It draws most heavily on the African cross-cultural literature and on my own background and clinical experience in South Africa. The answer is "yes" with regard to its method and attitude. I have striven for humility in the face of the different and imperfectly understood, and for modesty of interpretation. The assessment method I have outlined in Part III is anchored in my own experience, but in the hands of those who are themselves bearers of the native culture, or who are able to draw on indigenous interviewing and test administration resources, this cognitively based "behavioral neuropsychology" will in its present or modified forms be applicable in other settings.

## THE POLITICS OF CULTURE

The interplay of culture and cognition raises problems that are both intellectually and politically demanding. Intellectually, the neuropsychologist is compelled to confront and deal with the problem of construct validity, which as pointed out in the previous section is one of the most abstract and elusive in psychology (chap. 5). Politically, psychology and neuropsychology have always regarded themselves as above politics, which they can never be (Nell, 1997a). But the study of cultural difference very soon runs into questions about the values and attitudes that underlie the study of ability differences between people—and anything that suggests that there are intellectual disparities between ethnic or cultural groups at once comes under suspicion as potentially racist. Sometimes these suspicions are well-founded. The late Hendrik Verwoerd, who was a professor of applied psychology at the University of Stellenbosch, and subsequently a South African prime minister, was the architect of Grand Apartheid, the last (and longest lasting) of the 20th century's racist experiments; South Africa's attempt to clothe its racism in cultural fancy dress heightened the suspicion that cross-cultural psychology had been drafted to serve a system that used putative cultural differences to disenfranchise and harass its citizens. Psychologists opposed to these practices needed

> to avoid being regarded as, in any way, justifying the kinds of inequalities that are entrenched by law and custom in our society. There is a serious concern on the part of responsible researchers that their work should not feed into social systems that may pervert the meaning of the research and

> use its findings to justify repressive policies. This problem is particularly acute in psychology, as any suggestion of differences between groups can easily be twisted and made the basis for treating people differentially. (Miller, 1984, p. 4)

This "twisting" of group differences so as to find a justification for treating different people by different standards has a long and distasteful history in psychology. It would be pleasanter and save a lot of time if this history could be committed to psychology's trash can—but it cannot: The racist taint of IQ testing reflects directly on ability measurement, which is one of the professional foundations of clinical neuropsychology.

## EVOLUTION, RACISM, AND INTELLIGENCE TESTING

Darwin's *Origin of Species* (1859/1902) had an enormous impact on late 19th-century thought. One consequence was the development of a pseudo-scientific racism that had a comfortable fit with preexisting prejudice. These connections are brilliantly set out in Gould's (1981) account of racism as a forerunner of the IQ concept. His book, *The Mismeasure of Man,* won the National Book Critics' Circle Award and became a bestseller. Bestsellers speak to popular consciousness, and Gould's denunciation of the denial of hope by the reification of IQ measures was enormously popular.

In the 18th and 19th centuries, according to Gould, racists sought scientific respectability in the doctrines of the polygenists, who held the races to be descended from different Adams, so that "as another form of life, blacks need not participate in the equality of man" (Gould, 1981, p. 39). Polygeny became known as the "American School" of anthropology; Gould remarked that a nation practicing slavery and expelling Indians from their homelands would naturally incline to this view (p. 24). Monogenists, on the other hand, held all men to be descended from one Adam with "degenerationism" invoked to account for the development of different races.

In the colonies, civilization came to be equated with the extirpation of wilderness, so that what had been untamed could be brought to a garden-like state of ordered subjugation more acceptable to God (see Nash, 1967). At the same time, Whiteness and the values it stood for were to be preserved. Benjamin Franklin linked these themes by advocating both an environmental and an ethnic racism, calling for the elimination of America's woods—its indigenous forests—and its "tawneys":

> While we are, as I may call it, scouring our planet, by clearing America of woods, and so making this side of our globe reflect a brighter light to the eyes of the inhabitants in Mars or Venus, why should we . . . darken its people? Why increase the sons of Africa, by planting them in America where

> we have so fair an opportunity, by excluding all blacks and tawneys, of increasing the lovely white and red? (Benjamin Franklin, *Observations Concerning the Increase of Mankind,* 1751, cited in Gould, 1981, p. 32)

Darwin's concept of recapitulation encouraged museums to arrange displays illustrating the descent of man so that the lower apes were placed on the left, moving right to the higher primates next and then to the Negroes, followed by Asian peoples; that pinnacle of evolutionary achievement, the white European, stood to the farthest right: "Thus it happens that out of savages unable to count up to the number of their fingers and speaking a language containing only nouns and verbs, arise at length our Newtons and our Shakespeares" (Herbert Spencer, 1887, *Principles of Psychology,* cited in Cole & Scribner, 1974, p. 14). Gould remarked that both the doctrine of recapitulation and Lombroso's *stigmata* (physical marks of human criminality) were in essence "the search for signs of apish morphology in groups deemed undesirable" (1981, p. 113). In 1990, Kamin was the keynote speaker at the "Psychology and Apartheid" conference in Cape Town, where he first scandalized, then riveted, the audience by announcing that his topic for the evening was the size of the Black penis—and then going on to a virtuoso analysis of racist anthropometrics.

However, racist beliefs lie deep in human nature. In the first place, there is an innate human propensity to divide things into two—up and down, God and the devil, White and Black, truth and falsehood: In the human body, "up" is mind and wholesome food, "down" is lustfulness and the filth of excrement. For the White races, the coloring of up is light and white, and the coloring of down is dark and brown (i.e., the color of excrement and the nether orifice). As Manganyi brilliantly showed in "The Body for Others" (1991), this artefact of the white body has had tragic consequences in the association of evil with darkness and thus with the world's tawny people (for other examintions of the psychology of racism, see Mannoni's *Prospero and Caliban,* 1956; Joseph Conrad's *Heart of Darkness,* 1902/1960, in which compassion struggles against a deep-seated social Darwinism; and Dabydeen, 1985).

As a biologist, the force of Gould's book derives from the unexpected links he forged between biological notions of human origin and the drive to find a number that expresses human worth. He argued that the ordering of humanity on a supposedly evolutionary scale derives from the human propensity for *ranking,* that is, "for ordering complex variation as a gradual ascending scale" (1981, p. 24), as in Lovejoy's essay on the great chain of being (1936), or Bury's treatment of the idea of progress (1920; both cited in Gould, 1981). When combined with an equally beguiling fallacy, *reification,* which is the tendency to convert abstract concepts into concrete entities, the outcome is quantification, which embodies both the ranking and reification fallacies. The stage is thus set for the conversion of that most multifaceted and elusive of concepts, human ability, to a number, supposedly equivalent to one's global intelligence. People can now be ranked "in a single series of worthiness, invariably to find that oppressed and disadvantaged groups—races, classes, or sexes—are innately inferior and deserve their status" (p. 25).

As Gould's historical narrative moves to the 20th century, it is colored by the secondary source on which he relied—an angry, polemical, and influential book by Kamin, *The Science and Politics of IQ* (1974). The centerpiece of Kamin's argument was that the great triumph (and the great shame) of intelligence testing in the 1920s was the shaping of U.S. immigration policy to keep out Europe's feebleminded, inferior racial stock. Nathaniel Hirsch, who during these years was a research fellow in psychology at Harvard, quoted the following description with approval: "I have seen gatherings of the foreign born in which narrow and sloping foreheads were the rule. . . . In every face there was something wrong—lips thick, mouth coarse . . . sugarloaf heads . . . goose bill noses . . . a set of skew-molds discarded by the Creator" (in Kamin, 1974, p. 28).

This customary racism permeated early 20th-century lay and scientific thinking. One of the most telling passages in Gould's book draws on Kamin's account of how H. H. Goddard, Binet's American disciple, was drawn to Ellis Island, through which the huge influx of European migration to the United States was channeled. There he attempted to determine if Binet's new scale might serve to keep the feebleminded from setting foot on American shores. Kamin reported Goddard's findings as demonstrating that 83% of the Jews, 80% of the Hungarians, 79% of the Italians, and 87% of the Russians who were immigrants on Ellis Island were feebleminded. Gould (1981) took up the story in the following way:

> Consider a group of frightened men and women who speak no English and who have just endured an oceanic voyage in steerage. Most are poor and have never gone to school: many have never held a pencil or pen in their hand. They march off the boat; one of Goddard's intuitive women takes them aside shortly thereafter, sits them down, hands them a pencil, and asks them to produce on paper a figure shown to them a moment ago, but now withdrawn from their sight. Could their failure be a result of testing conditions, of weakness, fear, or confusion, rather than of innate stupidity? Goddard considered the possibility but rejected it. . . . What but stupidity could explain an inability to state more than 60 words, any words, in one's own language during 3 minutes? . . . Or ignorance of the date, or even month or year? . . . Since environment, either European or immediate, could not explain such abject failure, Goddard stated: "We cannot escape the general conclusion that these immigrants were of surprisingly low intelligence." (pp. 166–167)

As with all polemics, the Kamin–Gould account has done some molding of history to its own ends. Commenting on the alleged feeblemindedness of Jews, Hungarians, and others, Dorfman (1982) maintained that this was not Goddard's statement but Kamin's, and that Goddard in fact attributed no hereditary defect to these immigrants. In 1983, Snyderman and Herrnstein accused Kamin of presenting a misleading account of the history of intelligence testing in the United States on three grounds. In the first place, they charged that he misrepresents the American psychological establishment as uniformly supportive of Goddard and other early IQ researchers, when on the contrary it was often critical and environmentalist rather than racist in its interpretation of early test results. Second, they reexamined the Congressional Record for 1923 and 1924, concluding that in drafting the 1924 Immigration Act, "Congress took virtually no notice of intelligence testing" (p. 986). Third, they claimed that Kamin (and Gould in his footsteps) misrepresented Goddard's findings: In the published account of his work, Goddard stated that he had preselected his samples to exclude both the "obviously feebleminded" and the "obviously normal," and himself hedged his conclusion by saying that his figures do not relate to immigrants in general "or even of the special groups named" (citations from Snyderman & Herrnstein, 1983).

Despite this racist taint, the obvious utility of IQ testing for civilian and military selection led organized psychology in the United States to propose a mass testing program in 1917, at the height of World War I. In his presidential address to the American Psychological Association, Yerkes committed American psychology to assist the war effort. And this was not empty rhetoric. Yerkes (who was a major in the U.S. Army Sanitary Corps because, as Sarason, 1981, sarcastically remarked, he could not as a psychologist be admitted to the Medical Corps) led the U.S. Army testing program, applying the Army Alpha and Beta to 1,726,000 men, which was the largest testing program in history. Matarazzo (1972) noted that the Army Alpha included all 11 of the Wechsler–Bellevue subtests (with some also in the Army Beta, developed for non-English speakers and illiterates[1]), thus providing a huge laboratory for the new science of IQ testing.

## CLASS, RACE, AND CLINICAL NEUROPSYCHOLOGY

The two ghosts that continue to haunt neuropsychology both derive from psychology's historical links with Western colonialism. One is the universalist claim, already reviewed in this chapter and again in chapter 2, and

[1]Claassen (1997) recorded that in South Africa, M. L. Fick applied the Army Beta to African, "Colored," Indian, and "poor White" schoolchildren (in today's terminology, these would be the lower ranks of the working class). He published his results in 1929.

the other is an unconscious acceptance of the class and race divisions that continue to determine access to health care resources. In the developing countries, the former colonial health care machinery, including a more or less embryonic psychology, has passed into the hands of a new, indigenous elite, so that access and service quality are still differentially determined by class status. It is at the same time increasingly evident that many international development agencies have vested interests in preserving the power and wealth of establishment elites at the expense of the social underclass (Nell, 1996; Werner & Sanders, 1996).

The fundamental act of resource allocation is political, whether this is at the level of state decisions about spending, or at the personal level of seeking out a career in underserved areas (Andersson & Marks, 1989). Like other health care workers, neuropsychologists are enmeshed within a political system that continues to ensure that the educated, the articulate, and the male (boy babies are routinely better fed than girls wherever poverty is endemic; Bennet & George, 1987) have a better bite at the health care cake than the poor, the ill-educated, and women.

Those at the bottom of the social heap—in inner-city slums, the *favellas* of Brazil, and in the new military dictatorships all over the world that seek nothing but the enrichment of the powerful—are consumed by hopelessness and internalized violence (Bulhan, 1985; Nell & Brown, 1990). Among the poor of every country, the disabled are the poorest, and the mentally ill are the poorest of the poor; persons with significant brain damage are doubly disadvantaged because they are often labeled as both "crippled" and "crazy."

In these situations of oppression, psychologists have a uniquely important role. Alone among the health care professions, psychology cannot tolerate a passive client. Psychology's imperative demand of its clients is to mobilize their own resources and to heal themselves through their own strength. For all people, and especially for the social underclass, this injunction is politically important, because it constitutes a first and essential step toward the demand for full democratic participation both in health care and in wider political contexts.

In the developed countries of the West, state-funded social safety nets have over the past half century systematically attenuated these differences. But, in the emerging postindustrial economies of Western nations, harsher social policies are emerging that undermine hard-won gains. So it is not only in the "poor south" that practitioners need to be aware of the factors that shape access to and delivery of the basic commodities of food and health care.

Compassionate and politically sensitive neuropsychologists will avoid being overwhelmed by the bitter conjunction of being crippled and crazy that has already filled the client with helplessness. They will instead become case managers to mobilize welfare and personal resources around the client system that may be powerful enough to overcome these burdens. At the same time, they will ensure that they do not overwhelm those who have suffered brain damage by a review of symptoms that stresses their incurability, and a prognosis that offers truth untempered by hope.

Underlying these specific therapeutic injunctions is the need to remain sensitive to the disparities in power and status between professionals and those they serve.

## HONORING DIFFERENCE, RESPECTING EQUALITY

Despite episodes of political misuse, cross-cultural psychology is uniquely able to provide a humane framework within which the best interests of minority and culturally different peoples can be served by recognizing culture-specific differences rather than by an ideological universalism that seeks to obliterate these differences. Suppressing cross-cultural research and applications, however politically correct this may appear to be at a given historical moment, does a disservice to human welfare (see, e.g., chap. 3, "Cultural Difference and the Suppression of Science"). Without cross-cultural psychology, free rein is given to claims made under the banner of universalism that a test or a test battery is "culture-free"—a description born in the never-never land of psychometric wishful thinking that at once brands the claim as spurious.

On the contrary: the route to a universalist neuropsychology lies through local particulars. The wily emics lurk in the indigenous etic woodlands (see Footnote 1, p. 86), and will be flushed out of hiding by indigenous researchers launching their own substantial score-gathering projects for local populations on core neuropsychological tests, with standardized quantitative methods used to report the principal moderator variables of the test scores of culturally different clients. These are education, language proficiency, and test-wiseness, as well as the stimulus value of the environment (of which urbanization is one aspect). These scores can then be used to establish the construct and predictive validity of the tests, and finally for the production of valid norms. One result of this process will be to end the damaging practice of overcompensation for performance differences through the use of artificially low norms (Melendez, 1994; Nell, 1997a) that make even significantly impaired individuals appear to be "normal."

Chapter 2

# The Failure of Universalism: Test Score Differences Across Countries and Cultures

*A spectre haunts human thought: If truth has many faces, then not one of them deserves trust or respect. Happily, there is a remedy: human universals. They are the Holy Water with which the spectre can be exorcized. But of course, before we can use human universals to dispel the threat of cognitive anarchy, which would otherwise engulf us, we must first* find *them. And so, the new hunt for the Holy Grail is on.*

(Gellner, 1981, p. 1)

If mind, like brain, is one, and therefore unitary in all humans, then neuropsychological assessment founded on human universals will work equally well in London, New York, or the subsistence farming villages of South Africa and Brazil. If mind is many, however, and the ways in which people think and solve problems are determined by the interaction of their genetic endowment and the material conditions of their culture, then identical tests may make geniuses of average people in one culture and imbeciles of equally average people in another.

## THE UNIVERSALIST CANON

All psychological assessment faces these problems of culture and cognition, but they are especially acute for neuropsychology. Neuropsychologists are trained in hospital settings, in which the universality of physical diagnosis is unchallenged. The presence or absence of a pathogen in the bloodstream, nystagmus, vibration sense, or the elicitation of a reflex are all independent of culture or education. Physicians trained in England or North America can diagnose and treat with equal accuracy in Africa or Asia because at the level of these physiological functions, there are indeed species-wide normative expectations.

But one cannot conflate physiological with cognitive universals—although mainstream neuropsychology has consistently done so. Ward Halstead (1947) believed that his impairment index tapped into "biological intelligence," and his disciple Ralph Reitan held to this universalist belief. Reitan and Davison (1974) commented on what they interpreted as the convergence of results from cross-cultural studies:

> [This] fosters considerable confidence that certain brain–behavior relations are quite fundamental to the human organism regardless of cultural influence, and that the behavioral variance due to these organic factors is potent enough to filter through international barriers to communication. (pp. 11–12)

Difficulties of a particular kind arise when the specific functions that express these universal abilities must be listed. In both the second and third editions of *Neuropsychological Assessment* (1983, 1995), probably the most-cited and authoritative text in modern clinical neuropsychology, Lezak took the view that "the norms for some pschological functions and traits serve as species-wide performance expectations for adults" (1995, p. 99). As examples, she named motor and visuomotor control and coordination, speech, perceptual discrimination, orientation to space, binaural hearing, fine tactile discrimination, and discrimination between noxious and pleasant stimuli. However,

> some other skills that almost all physically intact adults can perform are counting change, drawing a recognizable person, basic map reading, and using a hammer and saw or basic cooking utensils. Each of these skills . . . is sufficiently simple that its mastery or potential mastery is taken for granted. Anything less than an acceptable performance in an adult raises the suspicion of impairment. (Lezak, 1995, p. 100)

This is a surprising list even for a dedicated universalist. On the one hand, it names physiological universals that people share not only with their species, *homo sapiens*, but with all primates and, in some cases, with the entire animal kingdom. On the other hand, it includes complex psychological functions such as map reading and drawing a person. Here in a nutshell is the radical universalist expectation: If an adult is unable to turn in an acceptable performance on a task judged to be a human universal on the basis of its appearance in one culture, such as change counting or map reading, then the impairment hypothesis arises, even though that individual may be the bearer of a culture in which coinage is not used and maps unknown; on drawing a person, considerable differences between cultures have repeatedly been found (Freeman, 1984; Richter, Griesel, & Wortley, 1988, in South Africa).

## THE ETERNAL DEBATE

It is helpful to place this universalist expectation in neuropsychology within the context of a much older debate among philosophers and anthropologists about rationality and relativism (Claxton, 1988; Gellatly, Rogers, & Sloboda, 1989; Hollis & Lukes, 1982; Lloyd & Gay, 1981): Do the commonalities indubitably shared by all humans (see chap. 4) override the different conditions under which individuals are born and mature, and the manifestly different regulatory mechanisms of human societies in different parts of the world? Or, do these external sources of variance on the contrary change the nature of human abilities in fundamental ways?

The universalist view is attractive because it frees psychology from the need to examine construct validity. If universals reign, then at worst one would need to establish new norms for existing tests; at best, even this would be unnecessary. However, universalism is also dangerous, because in the clash between "nativists" and "environmentalists" in the current debate about the origins of intelligence (Herrnstein & Murray, 1994), universalism leads directly to nativism: When the test performances of groups differ, this difference is attributed not to culture and environment—factors extrinsic to the individual—but to genetic endowment, an intrinsic or "native"[1] factor. The tension between nativists and environmentalists is a variant of the eternal debate between mentalists-idealists, who tend to universalism, and materialists-behaviorists, who prefer relativism.[2] Paradoxically, materialism-relativism (and therefore the expectation that wide variations in normative performances will occur) is the more tolerant of the two belief systems. Barnes and Bloor (1982) put this elegantly:

---

[1]The meaning of "native" and "nativist" lies in their origins, from the medieval Latin *nativus*, meaning produced by birth and therefore innate, as in Chaucer in the 14th century, "So angelic was her native beayty/ That like thing immortal seemed she." By extension, "native" has come to mean a person born in a particular place and having rights there by virtue of birth.

[2]Behaviorism is another and equally radical form of environmentalism: "Give me a dozen healthy infants, well-formed, and my own specified world to bring them up in and I'll guarantee to take any one at random and train him to become any type of specialist I might select—doctor, lawyer, artist . . . beggar-man and thief, regardless of his talents, penchants, tendencies, abilities, vocations and race of his ancestors" (J. B. Watson, *Behaviorism*, 1930, cited in Kamin, 1974, p. 178. Watson added that he is to be allowed to specify the type of world the children are to live in).

> Far from being a threat to the scientific understanding of forms of knowledge, relativism is required by it. Our claim is that relativism is essential to all those disciplines, such as anthropology, sociology, the history of institutions and ideas, and even cognitive pschology, which account for the diversity of systems of knowledge, their distribution and the manner of their challenge. It is those who oppose relativism, and who grant certain forms of knowledge a privileged status, who pose the real threat to scientific understanding of knowledge and cognition. (pp. 21–22)

In part, this is because relativism lends itself less readily than universalism to adoption by True Believers, of whom there has been no lack in psychology. Irvine and Berry (1988b) remarked that the survival of what they called the "fundamentalist canon," namely, a belief in universals of human performance, can be understood only if one takes into account the "high emotionality" of those who take this view: "It has to be seen as a sectarian belief system, somewhat authoritarian in its stance" (p. 53).

### The Universalist Expectation

In some settings, however, universalism seemed to be the only alternative to institutional racism. In South Africa, a country disfigured by Apartheid, the universalist hope had a poignant appeal that is ubiquitous in the writing of South African psychologists, for example:

> Separate norms for one group of people as opposed to another might be seen as a manifestation of the separatism that divides our lives. . . . We anticipate that these norm tables will become outdated, or at least we hope that a changing society will ultimately make them unnecessary. In the meantime, they remove another excuse for why so little has been done for so many children. (Richter & Griesel, 1988, pp. ii–iii)

And, setting off from Johannesburg in 1958 to apply psychological tests to the San people of the Kalahari desert, Reuning (1988) wrote of himself and his colleagues:

> We were at that time convinced, perhaps somewhat naively, that basic mental processes are "universal."[3] . . . If this were not so, the talk of "one mankind" and of "equal opportunities for all" would not make sense. It must be possible, we thought, to demonstrate this by suitable experiments, including cognitive performance tests. Intelligence, too, as far as it is not preconditioned by particular beliefs and traditions, or programmed in a language very different

[3]Twenty years later, with unintended irony, Scarr (1978) wrote that "there is no IQ test that will show our common humanity because each is bound to the culture from which the particular sample of knowledge and skills is drawn" (p. 332).

> from ours, should have a similar basic structure, although not necessarily developed in every facet to the same performance level. (p. 464)

Within these careful limits, the universalist expectation is reasonable; but in a clinical context, these delicacies are soon swept away by reality demands.

In the mid-1980s, newly returned from the United States and still starry-eyed at the clarity and decisiveness of neuropsychodiagnosis in America, I was presenting weekly neuropsychology rounds at a mental hospital near Johannesburg. One of my clients was a 37-year-old African woman admitted for observation. She had 3 years of formal education, had been a dressmaker, and presented at interview as a bright and lively person, conversing coherently and responding with insight and sometimes sharpness to questions. During testing, at which a Black psychology graduate acted as translator, strange things began happening. On Raven's Colored Matrices, her answers were consistently and inexplicably wrong. The client's task on this test is to say which of six design alternatives is the best match for a gap of the same size and shape in the pattern printed above. But even on the easiest items, this client's responses were bizarrely incorrect. I began formulating hypotheses about a focal right parietal lesion, and asked her to copy the Greek cross from the Geometric Design Reproduction test of my screening procedure. She drew both horizontal arms on the right of the figure, and she could not draw a circle and square touching one another.

At the ward round that afternoon, I asked her to draw the Greek cross on the blackboard, expecting to demonstrate to my attentive colleagues this remarkable form of apractognosia, but, on this occasion, her performance was faultless! Later that day, I spent another hour with her, determined to puzzle out the source of the inconsistency. She was now able to draw any number of perfect Greek crosses, and her previously chaotic planning on the Complex Figure of Rey had become more orderly after several rehearsals—an object lesson to a beginning cross-culturalist in the importance of practice for illiterate and semiliterate clients (see chaps. 9 and 10). The right hemisphere hypothesis was crumbling—but what of the bizarre responses on Raven's matrices? Through the translator, I again asked her to select the best alternative, and then put the question I should have put before: "Why?" She readily explained that she chose her response not because it matched the pattern (that seemed to her to be too easy) but because it made the most colorful and aesthetically pleasing patch on what she took to be a sheet of fabric with a piece torn out of it. In her own terms, as a dressmaker, her bizarre answers had been perfectly reasonable—although her score on the Ravens was of course in the defective range.

If as mainstream neuropsychology holds there are universal performance expectations, then the dressmaker's performance on the Ravens would have labeled her as severely impaired, despite clear subsequent evidence that her underlying abilities were intact. That afternoon, the seeds of doubt were planted; despite the seduction of universals, relativism might be a more humane and intellectually correct approach to cross-cultural neuropsychology. Universalism's failure to account for test score differences across countries and cultures is demonstrated by the data reviewed next.

## THE WORLD HEALTH ORGANIZATION NEUROBEHAVIORAL CORE TEST BATTERY

A large comparative database has been accumulated for the World Health Organization Neurobehavioral Core Test Battery (WHO–NCTB), set out in a series of recent publications (Anger et al., 1993; Brown, Wills, Yousefi, & Nell, 1991; Cassitto, Camerino, Hanninen, & Anger, 1990; Chia, Jeyaratnam, Ong, Ng, & Lee, 1994; Escalona, Yanes, Feo, & Maizlish, 1995; Lee & Lee, 1993; Liang, Chen, Sun, Fang, & Yu, 1990; Maizlish, Parra, & Feo, 1995; Nell, Myers, Colvin, & Rees, 1993; Richter et al., 1992; Rosenstock, Keifer, Daniell, McConnell, & Claypoole, 1991; Tang et al., 1995) and in unpublished sources (Anderson & Macpherson, in draft; Makunga, 1988; Nell, Kruger, Taylor, Myers, & London, 1995; Sesel, 1990).

These data are summarized in Table 2.1, in which the columns have been numbered for ease of reference. The table reports scores for 24 studies in 13 countries on 4 continents. All are for healthy subjects and, at least for the South Africans, exclude those with any previous history of head injury or psychiatric illness. This is the largest comparative international database available; although in many respects it falls far short of acceptable psychometric standards, which (as discussed in chap. 5) is a pervasive problem in neuropsychology. Sample sizes are often very small, as in columns 1–5, 9, and 14–16. Some results arouse suspicion, as for example the very low Aiming Correct score for Italy (col. 6), with an unduly high standard deviation. Education, which (see chap. 4) is the single most powerful moderator of test performance, is inadequately specified. For Nicaragua, we are told only that 12 of the 36 subjects had no formal education, with no data provided for the remaining 24. It may therefore be that the quite marked age declines seen for Austria, France, the Netherlands, and Poland are mediated by education rather than age, because education may covary negatively with age, as reported in an earlier study (Nell et al., 1993).

An urgent need in cross-cultural neuropsychological research is not only that the same core marker tests are used cross-nationally (a need partly met by the WHO–NCTB), but that variables are consistently reported using a standardized reporting format (chap. 11).

Six of the seven WHO–NCTB tests are reported, omitting the Profile of Mood States (POMS) because it is not valid for people from other

TABLE 2.1
Scores on the WHO–NCTB for 20 Studies in 13 Countries

| *Continent* | | *Europe* | | | | | | | | | |
|---|---|---|---|---|---|---|---|---|---|---|---|
| *Column* | | *1* | *2* | *3* | *4* | *5* | *6* | *7* | *8* | *9* | *10* |
| *Country/City* | | *Austria*[1] | | *France*[1] | | *Hungary*[1] | *Italy*[1] | *Netherlands*[1] | | *Poland*[1] | |
| *N* | | 19 | 15 | 24 | 22 | 18 | 112 | 35 | 41 | 28 | 25 |
| Age (*SD*)[2] | | 30 | 40 | 30 | 40 | 30 | 30 | 30 | 40 | 30 | 40 |
| Education Yrs (*SD*) | | 8–10 yrs schooling | | | | | | | | | |
| Reaction | *M* | 225 | 260 | 220 | 225 | — | 265 | 231 | 242 | 247 | 254 |
| Time | *SD* | 38.3 | 31.5 | 13.5 | 28.3 | — | 51.8 | 22.7 | 30.1 | 19.1 | 22.6 |
| *Santa Ana* | | | | | | | | | | | |
| Dom Hand | *M* | 49.9[3] | 44.3 | 50.9 | 46.4 | — | 43.6 | 43.8 | 43.8 | 42.6 | 39.7 |
| | *SD* | 6.6 | 6.7 | 5.7 | 6.4 | — | 6.0 | 6.0 | 4.9 | 4.6 | 5.2 |
| NonDom | *M* | 44.8 | 41.9 | 47.1 | 44.5 | — | 39.4 | 43.6 | 42.7 | 41.1 | 36.3 |
| Hand | *SD* | 5.6 | 6.4 | 0.9 | 4.9 | — | 5.7 | 5.4 | 4.6 | 4.1 | 4.5 |
| Aiming | | | | | | | | | | | |
| Correct | *M* | 143 | 152 | 174 | 177 | 181 | 88 | 155 | 150 | 240 | 212 |
| | *SD* | 35.9 | 51.3 | 39.9 | 51.0 | 30.2 | 55.3 | 28.9 | 33.8 | 31.7 | 46.6 |
| Total | *M* | 173 | 175 | 175 | 179 | 200 | 152 | — | — | 246 | 222 |
| | *SD* | 38.6 | 50.4 | 40.5 | 51.8 | 28.5 | 44.3 | — | — | 34.7 | 55.7 |
| Digit Span | | | | | | | | | | | |
| Forward | *M* | 7.7 | 8.4 | 6.6 | 5.7 | 6.1 | 6.7 | 6.1 | 6.4 | 5.4 | 4.0 |
| | *SD* | 1.9 | 2.0 | 2.0 | 1.9 | 0.9 | 1.9 | 1.7 | 1.8 | 1.1 | 0.7 |
| Backward | *M* | 6.7 | 7.9 | 5.8 | 4.7 | 4.5 | 4.5 | 6.4 | 6.1 | 5.0 | 3.7 |
| | *SD* | 1.3 | 3.1 | 2.3 | 1.9 | 0.9 | 2.1 | 1.9 | 1.9 | 0.9 | 0.8 |
| Digit | | | | | | | | | | | |
| Symbol | *M* | 49.9 | 51.6 | 55.3 | 50.5 | 46.5 | 42.3 | 55.1 | 53.8 | 53.3 | 46.4 |
| | *SD* | 9.0 | 12.5 | 9.8[4] | 10.3 | 9.2 | 10.6 | 11.9 | 10.9 | 10.6 | 11.1 |
| Benton | | | | | | | | | | | |
| Visual | *M* | 9.1 | 9.2 | 9.8 | 8.8 | — | 8.2 | 8.8 | 8.6 | 8.5 | 8.3 |
| Retention | *SD* | 0.6 | 0.8 | 1.1 | 1.2 | — | 1.6 | 1.1 | 1.1 | 1.2 | 1.6 |

*(Continued)*

TABLE 2.1
*(Continued)*

| Continent | | Asia | | | | | South America | | Africa | | |
|---|---|---|---|---|---|---|---|---|---|---|---|
| Column | | 11 | 12 | 13 | 14 | 15 | 16 | 17 | 18 | 19 | 20 |
| Country/City | | Korea[5] | China[6] | | China[7] | Singapore[8] | Nicaragua[10] | Venezuela[12] | Johannesburg[13] | Johannesburg & Durban[14] | Ceres[15] |
| N | | 81 | 49 | 19 | 23 | 21 | 36 | 47 | 54 | 228 | 247 |
| Age (SD)[2] | | 34.7 | 30 | 40 | 35.4 | 50.5 | 27.8 | 35 | 45.7 | 46.0 | 36.9 |
| | | (8.6) | | | (8.4) | (6.4) | (9.3) | (11) | (9.8) | (9.0) | (10.0) |
| Education Yrs (SD) | | 12.9 | Not given | | 6.2 | 4.9 | —[11] | 8 | 6.0 | 6.1 | 5.1 |
| | | (2.5) | | | (2.7) | (2.1) | | (3) | (3.1) | (3.5) | (2.9) |
| Reaction | M | 257 | 248 | 251 | 260 | — | 308 | 313 | 286 | 275 | 289 |
| Time | SD | 32.9 | 23.3 | 22.7 | 30 | — | 50 | 66 | 56.2 | 68.1 | 69.6 |
| *Santa Ana* | | | | | | | | | | | |
| Dom Hand | M | 45.6 | 40.5 | 39.3 | 37.7 | 16.7 | 35.6 | 42.6 | 34.4 | 35.9 | 38.4 |
| | SD | 5.1 | 4.9 | 4.6 | 7.3 | 0.04[9] | 7.0 | 7 | 3.7 | 7.7 | 6.7 |
| Nondom | M | 42.7 | 37.9 | 34.5 | 36.1 | 17.0 | — | 37.4 | 30.4 | 32.5 | 35.4 |
| Hand | SD | 5.4 | 5.3 | 3.9 | 5.3 | 0.04[9] | — | 7 | 3.4 | 7.0 | 7.7 |
| Aiming | | | | | | | | | | | |
| Correct | M | — | 235 | 210 | 147 | 154 | 94 | — | 81.6 | 79.9 | 91.6 |
| | SD | — | 31.2 | 18.7 | 44.0 | 0.06[9] | 29.9 | — | 18.3 | 36.6 | 34.9 |
| Total | M | — | 245 | 225 | 194 | 156 | — | 144 | 163.4 | — | 114.4 |
| | SD | — | 33.8 | 28.2 | — | 0.06[9] | — | 50 | 19.2 | — | 31.6 |

*(Continued)*

TABLE 2.1
*(Continued)*

| *Continent* | | *Asia* | | | | | *South America* | | *Africa* | | |
|---|---|---|---|---|---|---|---|---|---|---|---|
| *Column* | | *11* | *12* | *13* | *14* | *15* | *16* | *17* | *18* | *19* | *20* |
| *Country/City* | | *Korea*[5] | *China*[6] | | *China*[7] | *Singapore*[8] | *Nicaragua*[10] | *Venezuela*[12] | *Johannesburg*[13] | *Johannesburg & Durban*[14] | *Ceres*[15] |
| Digit Span | | | | | | | | | | | |
| Forward | *M* | — | — | — | 6.8 | — | — | 5.2 | 5.2 | 5.2 | 4.8 |
| | *SD* | — | — | — | 3.5 | — | — | 2 | 1.0 | 1.0 | 1.0 |
| Backward | *M* | — | — | — | 4.6 | — | — | 4.3 | 3.3 | — | 3.3 |
| | *SD* | — | — | — | 1.6 | — | — | 2 | 1.0 | — | 1.1 |
| Digit | | | | | | | | | | | |
| Symbol | *M* | 59.0 | 55.6 | 52.7 | 39.6 | 38.0 | 25.4 | 37.1 | 22.1 | 20.9 | 25.4 |
| | *SD* | 13.6 | 9.4 | 9.3 | 11.8 | 0.07[9] | 11.9 | 14 | 10.1 | 9.5 | 10.4 |
| Benton | | | | | | | | | | | |
| Visual | *M* | 8.1 | 9.0 | 8.8 | 6.4 | 7.7 | 6.1 | 6.8 | 7.0 | 6.4 | 6.9 |
| Retention | *SD* | 1.5 | 1.0 | 1.1 | 2.3 | 0.07[9] | 2.2 | 2 | 1.6 | 1.7 | 2.0 |

[1]Cassitto et al. (1990)
[2]The designations 30 & 40 indicate subjects aged 26–35 and 36–45, respectively.
[3]Reported elsewhere in the table as 46.9 with the same *SD*.
[4]Reported elsewhere in the table as 9.08 with the same *SD*.
[5]Lee & Lee (1993)
[6]Liang et al. (1990)
[7]Tang et al. (1995)
[8]Chia et al. (1994)
[9]Standard error of measurement
[10]Rosenstock et al. (1991)
[11]Twelve of the 36 subjects are reported to have had no formal education; no data are given for the remaining 24. The mean number of years of education of these agricultural workers from a remote agricultural region is unlikely to be more than 3 or 4.
[12]Maizlish et al. (1995)
[13]Nell et al. (1993)
[14]Myers et al., unpublished data
[15]London, Myers, Nell, Taylor, & Thompson (1997)

cultures (Anger et al., 1993; Liang et al., 1990; Mokhuane, 1997; Nell et al., 1993; see chap. 4). All scores are for indigenous rather than settler populations (for some thought-provoking data on White South Africans, see "A Case Study as a Footnote" in chap. 4). Tests are more fully described in chapter 10, with only minimum detail here.

*Simple Reaction Time* (SRT) was measured using the Hick Box (Hick, 1952; Jensen, 1988), a more sophisticated device than the portable Terry Reaction Time Tester recommended for field use in the WHO–NCTB *Operational Guide.* The Hick Box records movement (or decision) time separately from reaction time. However, the comparability of reaction times measured by different devices is unclear, and this caution should be kept in mind when evaluating the data in the following tables.

Table 2.1 shows that the SRT ranges in milliseconds for Europe are 220–265 ms; when comparing the fastest times of 220 ms and 248 ms for the European and Asian samples, respectively, the latter are one standard deviation[4] slower (the mean standard deviation for both Europe and Asia is 29) but only about one third of a standard deviation slower if the mean times for the five Asian and six younger European groups are compared (243 ms as against 254 ms). The two South American countries are, on average, 2.3 standard deviations slower than the European countries. The South African workers are markedly faster than the South American samples at a mean of 282 ms, just 1.3 standard deviations slower than the European groups.

On the *Santa Ana Manual Dexterity Test,* the mean standard deviation on the dominant hand for the younger European groups is just under 6; the table shows that the South American and Asian groups (leaving the deviant Singapore score out of account) are about one standard deviation slower than the Europeans; the South Africans are 1.7 standard deviations slower.

*The Pursuit Aiming Test* is a typical dotting or targeting test (Carroll, 1993, p. 536). The mean total items attempted for Europe is 189, with a standard deviation of 37; surprisingly, given the very low educational level of the groups in columns 15 and 16 and this test's strong classroom component, the Asian mean at 198 is somewhat higher. However, the South American and South African groups are more than a standard deviation below the mean scores for Europe and Asia.

## Digit Span

This and Digit Symbol Substitution are subtests of the Wechsler Adult Intelligence Scale (WAIS) and the WAIS–R, the most widely used of all

[4]The standard deviation is referenced to the European samples in the discussion that follows.

intelligence tests. Using the 1981 WAIS–R norms (Wechsler, 1981), it is possible to assign standard scores[5] for Digit Span—derived by combining the scores for Digits Forward and Digits Backward—once the raw score has been calculated.[6] The WHO–NCTB data in Table 2.1 are not readily interpretable because they reflect only one score for the Asian and the South American samples. Table 2.2 sets out the additional available cross-national data on Digit Span, combining these for ease of analysis with the Digit Span scores in Table 2.1.

Using the span-to-score conversion method described in footnote 6, Table 2.2 shows that for the six European samples, standard scores range from 13 in Austria (a standard deviation above the U.S. mean) to 6 in Poland (1.3 standard deviations below the mean), with a mean standard score of 9.2. Wechsler's U.S. norms are now 50 years old, and although still cited (e.g., in Spreen & Strauss, 1991), they are now misleading. Taken together, the European and Autralasian data show that with the exception of Poland, these working-class groups score at or near the U.S. mean.

Turning to the southern hemisphere, in South America the working-class group (col. 15) is nearly 2 standard deviations below the U.S. mean, and more than 12 years education (col. 14) does not override whatever other effects may be present in the middle-aged Colombian group. In South Africa, on the other hand, the well-educated subjects in columns 16–18 do very well, with standard scores of between 8 and 12; however, the worker groups in columns 19–22 are uniformly 2 or more standard deviations below the U.S. mean.

*Digit Symbol Substitution* draws on two classroom-type skills: fine motor control and rapid learning of associates. On these grounds, one would anticipate that both it and Pursuit Aiming would show especially strong education effects. For the European samples, the WAIS–R standard score equivalents are between 6 and 9 with a mean of 8, and from 6 to 10 with a mean of 7 for the Asian groups, using the norm tables for 25- to 34-year-olds. In both cases this is a nonsignificant difference. For the South American and South African groups, using the 35- to 44-year-old norms, the standard score means are, respectively, 6 and 5—two thirds to a full standard deviation below the mean standard scores for the two former areas.

[5]With a mean of 10 and a standard deviation of 3.

[6]In order to estimate scores for this test, which are not given in the source data for Table 2.2, the convention adopted was that two points were given for each trial up to the last two trials, which scored only one point each. Thus, for a Digits Forward mean of 7.7, the span is rounded to 8, and 2 points given for each of the span lengths between 3 and 6, with 1 point for each of the last two trials (7 and 8), for a total of 10; to this is added the Digits Backward score, which for a mean of 6.7, rounded to 7, is also 10 (bearing in mind that the backward series begins with a span of 2). Using the norm tables for ages 25–34, a score of 20 is equivalent to a standard store of 13.

TABLE 2.2
Digit Span Scores for 21 Studies in 12 Countries

| *Continent* | *Europe* | | | | | | *N. America* | | *Australasia* | | |
|---|---|---|---|---|---|---|---|---|---|---|---|
| *Column* | *1* | *2* | *3* | *4* | *5* | *6* | *7* | *8* | *9* | *10* | *11* |
| *Country/City* | *Austria*[1] | *France*[1] | *Hungary*[1] | *Italy*[1] | *Netherlands*[1] | *Poland*[1] | *USA*[2] | *USA*[2] | *Australia*[3] | *China*[4] | *Israel*[5] |
| *N* | 19 | 24 | 18 | 112 | 35 | 28 | 50 | 46 | 44 | 23 | 22 |
| Age | Age 26–35 | | | | | | 20–29 | 40–49 | 16–18 | 35.4 (8.4) | |
| Years Education (*SD*) | Schooling 8–10 years | | | | | | —[a] | —[a] | Scholars | 6.2 (2.7) | |
| Forward Span | 7.7 | 6.6 | 6.1 | 6.7 | 6.1 | 5.4 | 7.0 | 6.0 | 6.7 | 6.8 | |
| *SD* | 1.9 | 2.0 | 0.9 | 1.9 | 1.7 | 1.1 | 1.2 | 1.1 | 1.2 | 3.5 | |
| Backward Span | 6.7 | 5.8 | 4.5 | 4.5 | 6.4 | 5.0 | 5.3 | 4.3 | 5.0 | 4.6 | |
| *SD* | 1.3 | 2.3 | 0.9 | 2.1 | 1.9 | 0.9 | 1.1 | 1.1 | 1.2 | 1.6 | |
| Score | 20 | 16 | 12 | 14 | 14 | 10 | 14 | 10 | 14 | 14 | |
| Standard Score[c] | 13 | 10 | 8 | 9 | 9 | 6 | 9 | 7 | 9 | 9 | 10 |

*(Continued)*

TABLE 2.2
*(Continued)*

| *Continent* | *South America* | | | *South Africa* | | | | | | |
|---|---|---|---|---|---|---|---|---|---|---|
| *Column* | *12* | *13* | *14* | *15* | *16* | *17* | *18* | *19* | *20* | *21* |
| *Country/City* | *Columbia*[6] | | *Venezuela*[7] | *SA 1*[8] | *SA 2*[9] | *SA 3*[9] | *SA 4*[10] | *SA 5*[11] | *SA 7*[12] | *SA 8*[13] |
| *N* | 43[b] | 43[b] | 47 | 100 | 140 | 63 | 247 | 54 | 15 | 20 |
| Age (*SD*) | 56–60 | | 35 | 24 | 25 | | 37 | 46 | 35–50 | 41 |
| | | | (11) | (4.2) | (3.3) | | (10.0) | (9.0) | | (8.2) |
| Years Education | 6–12 | >1 | 8 | 13–14 | 9–12 | 13–15 | 5 | 6 | 6 | 5 |
| | | (2) | (3) | | | | (2.9) | (3.5) | (1.7) | (8.2) |
| Forward Span | 5.1 | 5.8 | 5.2 | 6.2 | 7.4 | 7.9 | 4.8 | 5.2 | 4.3 | 4.5 |
| *SD* | — | — | 2 | 0.9 | 2.3 | 2.0 | 1.0 | 1.0 | 1.7 | 0.9 |
| Backward Span | 3.2 | 3.9 | 4.3 | 5.0 | 5.6 | 6.3 | 3.3 | 3.3 | 2.7 | 2.8 |
| *SD* | — | — | 2 | 0.9 | 2.2 | 2.0 | 1.1 | 1.0 | 1.1 | 1.4 |
| Score | 6 | 10 | 8 | 12 | 16 | 18 | 6 | 6 | 4 | 6 |
| Standard Score[c] | 4 | 7 | 5 | 8 | 10 | 12 | 4 | 4 | 3 | 4 |

[1]Cassitto et al., 1990
[2]Wechsler, 1945
[3]Ivinskis, 1971
[4]Tang et al., 1995
[5]Richter et al., 1992
[6]Ardila et al., 1994
[7]Maizlish et al., 1995
[8]Makunga, 1988
[9]Avenant, 1988
[10]London et al., 1997
[11]Nell et al., 1993
[12]Sesel, 1990
[13]Brown et al., 1991

**Notes**
[a]Education not given
[b]With 346 subjects and 8 cells, estimated cell size is 43
[c]Norms for ages 25–34 used except Columns 8, 10, 18 and 19–21 (Ages 35–44), 9 (Ages 18–19), & 12–13 (Ages 55–64)

On the *Benton Visual Retention Test,* mean recall is 8.9 designs for the six younger European samples, with a mean standard deviation of 1.1. The mean is two standard deviations lower for the Asian samples, and three standard deviations lower for the South American and South African samples.

## WECHSLER INTELLIGENCE SCALES

Zindi (1994) reported the results of a comparison between 204 working-class Black Zimbabwean children and a matched sample of 202 White pupils in London on 10 of the 12 subtests in the Wechsler Intelligence Scale for Children–Revised (Table 2.3). On each of the subtests, the Zimbabwean children score about a standard deviation below the English group; their mean score is .85 of a standard deviation (2.56 scale points) lower. Across the 10 subtests, the English children have a total score of 95.3, while the Zimbabwean children's total is 69.7: Overall, the individual English subtest scores are up to one scale point (.3 of a standard deviation) below the WISC–R mean, while the Zimbabwean children are 3 to 4 scale points (1 to 1.3 standard deviations) lower.

Avenant (1988) reported the WAIS–R performances of 203 Black South Africans, of whom 140 were prison wardens with an education of between 9 and 12 years, and 63 undergraduate students at the all-Black universities[7] of Fort Hare, Zululand, the North, and the Medical University of South Africa. Mean age was 24.8 years (*SD* = 3.3). The test was "adapted" for administration in English by trained testers, with the wording of some items changed to prevent purely local difficulties in understanding (Avenant, 1988, pp. 3–4).

Table 2.4 sets out the raw and scaled scores obtained by each of these groups. The university undergraduates are one standard deviation below the U.S. age norms on the first three items of the verbal scale and on Block Design; nonsignificantly below the norm on Digit Span and Similarities; and rather more than a standard deviation below the norm on Picture Completion, Picture Arrangement, and Object Assembly. Their Full Scale IQ (FSIQ) is 1.3 standard deviations below the U.S. norm. The prison wardens, with their lower level of education, fare predictably worse, with an FSIQ 1.8 standard deviations below the U.S. norm. Raw score differences between the two groups are in all cases except for Digits Forward significant at better than the 5% level. However, the language artefact would have

[7]South Africa now has a democratically elected government, and these universities are now open to students of all races. However, the legacy of Apartheid education lives on, and it will take decades to eliminate its effects.

TABLE 2.3
Mean WISC–R Scores for Zimbabwean and English Pupils (Zindi, 1994)

| | *Zimbabwean* (N = *204*) | | *English* (N = *202*) | |
|---|---|---|---|---|
| *WISC–R Variable* | *M* | *SD* | *M* | *SD* |
| Information | 7.44 | 2.02 | 9.80 | 2.33 |
| Similarities | 7.23 | 2.25 | 9.47 | 2.87 |
| Arithmetic | 7.43 | 2.15 | 9.43 | 2.94 |
| Vocabulary | 6.48 | 2.27 | 9.42 | 2.27 |
| Comprehension | 6.27 | 2.13 | 9.53 | 2.23 |
| Picture Completion | 7.24 | 2.97 | 9.92 | 2.92 |
| Picture Arrangement | 8.01 | 2.79 | 9.87 | 2.47 |
| Block Design | 6.46 | 2.25 | 9.89 | 3.01 |
| Object Assembly | 6.93 | 2.41 | 8.98 | 2.25 |
| Coding | 6.24 | 3.03 | 8.96 | 2.49 |

*Note.* First published in *The Psychologist,* Bulletin of the British Psychological Society, December 7, 1994.

affected the scores of both groups to an unknown extent, with a proportionately greater effect on the scores of the lesser educated subjects.

Preliminary data gathered by the UNISA Health Psychology Unit for 157 Black South Africans with less than 12 years education, reported in Table 2.5, raise similar concerns. These competent men, all in long-standing employment in a sophisticated environment, score between about 1 and 2 standard deviations below the U.S. WAIS–R norms. There is no reason to believe that these individuals have a lower ability level than the age-matched U.S. and English norm groups with whom they are compared. The alternative hypothesis is that for whatever reason, the Wechsler tests lack validity for these subjects, and their use for such subjects would violate fundamental equity principles (i.e., the South African legislature's prohibition on the use of psychometric tests that are invalid or biased).

## LIST LEARNING AND PARAGRAPH RECALL

Anderson and Macpherson (in draft) gave norms for 163 Black schoolchildren from age 6 to 16 in the Pietermaritzburg area of South Africa's Natal province; the results were stratified by age, sex, and rural or urban school. All tests were administered in Zulu. Combining the urban and rural scores, Table 2.6 gives their norms across five trials for the 12- to 14-year-olds and 14- to 16-year-olds on the Rey Auditory Verbal Learning Test (AVLT), on which they administered the interference list, but not Trials 6 and 7. These scores can be compared with the results for like-aged groups reported by Geffen et al. (1990).

TABLE 2.4
Raw and Scaled Scores on the WAIS–R
for Two Groups of Black South Africans

| | *Education [Mean Age (Both Groups) 24,8 (3,3)]* | | | | | |
|---|---|---|---|---|---|---|
| | *9–12 Years (N = 140)* | | | *University Undergraduates (N = 63)* | | |
| *Name of Test* | *Raw* | *(SD)* | *Scaled** | *Raw* | *(SD)* | *Scaled** |
| *Verbal Scale* | | | | | | |
| Information | 9.7 | (4.0) | 6 | 14.4 | (3.3) | 7 |
| Comprehension | 10.9 | (2.9) | 6 | 14.8 | (3.0) | 7 |
| Arithmetic | 8.4 | (2.4) | 7 | 9.2 | (2.1) | 7 |
| Digit Span | | | 8 | | | 9 |
| Forward | 7.4 | (2.3) | | 7.9 | (2.0) | |
| Backward | 5.6 | (2.2) | | 6.3 | (2.0) | |
| Similarities | 14.1 | (3.8) | 7 | 18.6 | (3.9) | 9 |
| *Performance Scale* | | | | | | |
| Picture Comp | 9.8 | (3.4) | 5 | 11.4 | (3.0) | 6 |
| Object Assem | 18.9 | (7.3) | 5 | 21.3 | (6.6) | 6 |
| Block Design | 17.8 | (7.6) | 6 | 21.9 | (7.3) | 7 |
| Digit Symbol | 43.5 | (10.0) | 7 | 50.9 | (8.7) | 8 |
| Picture Arr | 5.4 | (3.3) | 6 | 7.3 | (3.7) | 6 |
| Verbal IQ | | | 76 | | | 80 |
| Performance IQ | | | 71 | | | 76 |
| Full Scale IQ | | | 73 | | | 77 |

*Scaled scores calculated from Avenant (1988, Table 6.3) and Wechsler (1981, Tables 19 and 20).

Reproduced by permission of the Human Sciences Research Council, Pretoria, South Africa.

TABLE 2.5
Scores of 157 Black South Africans with Less Than
12 Years Education on Six Psychological Tests

| *Variable* | N | *M* | *SD* |
|---|---|---|---|
| Age | 157 | 43.37 | 8.85 |
| Highest School Grade Passed (Grade 12 = Std 10) | 154 | 7.08 | 2.88 |
| WAIS–R Digit Symbol* | 152 | 6.32 | 1.97 |
| WAIS–R Block Design* | 156 | 6.26 | 1.71 |
| WAIS–R Digit Span: Forward Span | 155 | 4.74 | 1.06 |
| WAIS–R Digit Span: Backward Span | 155 | 3.36 | 0.87 |
| WAIS–R Digit Span: Standard Score* | 157 | 5.54 | 1.85 |
| WISC–III Arithmetic** | 153 | 4.98 | 2.13 |
| WISC–III Mazes** | 152 | 5.67 | 4.56 |

*Standard scores are age-matched.

**WISC–III standard scores all for age 16–11.

TABLE 2.6
Means (*SDs*) on the Rey Auditory Verbal Learning Test

| | *Anderson & Macpherson (in draft)* *Age 12–14* | | | | | | *Anderson & Macpherson (in draft)* *Age 14–16* | | | | | |
|---|---|---|---|---|---|---|---|---|---|---|---|---|
| | *Male* N = *10* | | *Female* N = *10* | | *Total* N = *20* | | *Male* N = *16* | | *Female* N = *17* | | *Total* N = *33* | |
| | *M* | *SD* | *M* | *SD* | *M* | *SD* | *M* | *SD* | *M* | *SD* | *M* | *SD* |
| Trial 1 | 5.0 | 1.41 | 5.5 | 0.97 | 5.3 | 1.21 | 5.9 | 0.89 | 6.1 | 1.48 | 6.0 | 1.21 |
| Trial 2 | 7.7 | 2.16 | 7.8 | 2.49 | 7.8 | 2.27 | 8.3 | 2.30 | 8.4 | 1.93 | 8.3 | 2.09 |
| Trial 3 | 9.2 | 1.81 | 9.5 | 2.17 | 9.4 | 1.95 | 9.9 | 1.53 | 9.4 | 1.73 | 9.7 | 1.63 |
| Trial 4 | 10.1 | 2.08 | 10.3 | 1.77 | 10.2 | 1.88 | 11.2 | 1.80 | 10.8 | 1.85 | 11.0 | 1.80 |
| Trial 5 | 12.0 | 2.06 | 12.0 | 1.49 | 12.0 | 1.75 | 11.8 | 2.18 | 10.8 | 1.75 | 11.2 | 2.0 |
| List B | 5.1 | 2.09 | 5.2 | 1.70 | 5.2 | 1.90 | 5.9 | 1.75 | 5.5 | 1.28 | 5.7 | 1.51 |
| Total Score (Trials 1–5) | 44.0 | | 45.1 | | 44.7 | | 47.1 | | 45.5 | | 46.2 | |

TABLE 2.7
Paragraph Recall on the Wechsler Memory Scale for Seven Studies

| *Country* | *United States*[1] | | | *South Africa* | | | |
|---|---|---|---|---|---|---|---|
| *N* | 74 | 67 | 41 | 100[2] | 20[3] | 20[4] | 54[5] |
| Age (*SD*) | 18–29 | 30–39 | 40–49 | 24 (4.2) | 41 (8.19) | 25 (7.2) | 45.7 (9.8) |
| Yrs Ed. (*SD*) | — | — | — | 13–14 | 4.9 (8.20) | 10 (1.8) | 6.0 (3.1) |
| *Immediate* | | | | | | | |
| Total Score | 22.9 | 24.6 | 23.4 | — | | | |
| *SD* | 6.7 | 7.0 | 5.0 | — | | | |
| Prorated | 11.5 | 12.3 | 11.7 | — | | | |
| One story | | | | 12.1 | 8.1 | 13.4 | 9.7 |
| *SD* | | | | 3.5 | 3.2 | 4.2 | 3.0 |
| *Delayed* | | | | | | | |
| Total | 19.8 | 22.2 | 21.1 | | | | |
| *SD* | 6.7 | 7.6 | 5.9 | | | | |
| Prorated | 9.9 | 11.1 | 10.6 | | | | |
| One story | | | | 11.6 | 7.4 | 14.2 | 8.6 |
| *SD* | | | | 3.8 | 3.2 | 4.9 | 3.4 |

[1]Abikoff et al. (1987)
[2]Makunga (1988); undergraduates
[3]Brown et al. (1991); unexposed manganese workers
[4]Adonisi (1988); students and clerical workers
[5]Nell et al. (1993); paint factory workers

Table 2.7 compares recent U.S. and South African norms on the illogically named Logical Memory task of the Wechsler Memory Scale. Abikoff et al. (1987) reported only a composite score for the two paragraphs combined: These scores have been halved to allow for comparison with the South African data, which used only one 22-item story, adapted from Wechsler's original. As with the Digit Span data reported in Table 2.2, the well-educated South Africans (cols. 4 and 6) do as well as or better than the Americans. However, the factory workers are about a standard deviation below their better educated compatriots. These results suggest that memory is a school-developed skill, and that despite the dependence of illiterate or semiliterate people on memory unreinforced by written reminders, narrative memory remains education dependent.

## MENTAL CONTROL AND FLUENCY

Psychologists in developing countries often make a plausible assumption that well-urbanized individuals who have been city dwellers for a generation or more, and thus constantly use numbers in their daily lives, will be able to count at a goodly pace, and, given the ubiquitousness of powerful tacit-knowledge number manipulation techniques (Sternberg, Wagner, Williams, & Horvath, 1995), will also be able to manipulate numbers quickly and with some fluency. However, as Irvine and Berry (1988b) pointed out in another context, "Theory does not support even in the most meaningless of tasks the assumption of stimulus equivalence" (pp. 51–52). More broadly, it cannot be taken for granted that the task-as-given and the task-as-received are in fact equivalent (Laboratory of Comparative Human Cognition, 1982). Such equivalence cannot be achieved by the simple substitution of diagrams and symbols for words and numbers (Kendall, Verster, & von Mollendorf, 1988), an approach much favored by the "culture-fair" and "culture-free" movements. This tradition arose in part from Cattell's attempt to address the problems of psychometric testing in other cultures by the production of culture-free tests, from which cultural influence would be entirely absent (supposedly so, e.g., for Raven's Progressive Matrices), or, more modestly, the "Culture-Fair Test of *g*" (Cattell, 1967; see also Carroll, 1984). Dague (1972) commented that "no test of intelligence could be independent of cultural factors" (p. 68).[8]

For example, it is taken for granted in Western society that adults will be able to count from 1 to 20 and to say the alphabet in little more than 5 seconds, and to count backward from 20 in 7 or 8 seconds. A group of 15 South African laborers with a mean of 6 years of education took on average 9.3 seconds ($SD = 4.0$) to count to 20, 25 seconds to count backward from 20, and 22.3 seconds ($SD = 14.6$) to say the alphabet.

Table 2.8 sets out the scores of South African and U.S. groups on controlled oral word generation and compares these with the norms of Yeudall, Fromm, Reddon, and Stefanyk (1986). The scores obtained by Makunga for South African undergraduates (row 3) are so low, despite their educational level, that one suspects that for whatever reason, these students at the University of Zululand chose not to cooperate with the researcher. The lesser educated South African group (row 2) are a standard deviation lower than the Americans (row 1).

An appropriate method to test verbal fluency in illiterate and semiliterate subjects is associate generation by category, given in row 4 for South African subjects, together with norms (row 5) for elderly Colombians from Ardila et al. (1994; further studies of this promising method are reviewed by Lezak, 1995, pp. 546–548). It is plausible to assume that level of education would make little difference, because the categories in question would appear to be equally accessible to all persons; however, as previously noted, plausibility is one thing and data are another—both the South African

[8]Dague echoed a perennial concern of those attempting to balance freedom from bias against predictive validity: He remarked that procedures based entirely on well-known local activities and skills would have had no predictive value for industrial applications, and quoted Fontaine's remark, "We are not selecting blacks to make bows and construct pirogues" (canoes) (Dague, 1972, p. 67).

TABLE 2.8
Controlled Oral Word Generation by Letter of the Alphabet and by Category in 60 Seconds (*SD*)

| | *Letter of the Alphabet* | | |
|---|---|---|---|
| *By Letter* | *First* | *Second* | *Third* |
| 1. Yeudal et al. (1986) (*N* = 225; 15–40 yrs; 14.9 yrs ed.) Letters F-A-S (English) | 15.2 (4.4) | 13.3 (4.1) | 16.7 (4.8) |
| 2. Moselenyane (1990) (*N* = 51, 25.3 yrs, 10.4 yrs ed.) Letters K-T-P (North Sotho) | 10.5 (3.4) | 11.3 (4.0) | 11.6 (4.0) |
| 3. Makunga (1988) (*N* = 100, 23.7 yrs, > 12 yrs ed.) Letters F-S-L (Zulu) | 4.8 (1.4) | 5.3 (1.5) | 5.4 (1.5) |
| *By Category* (*Sesel, 1990*) | *Animals* | *Clothing* | *Fruit & Vegetables* |
| 4. 15 Laborers, 6.1 (1.7) years education | 8.7 (2.5) | 9.1 (2.3) | 11.7 (9.8) |
| 5. 20 Professional South Africans, 17.6 (2.7) years education | 22.1 (5.2) | 21.5 (4.6) | 23.2 (5.4) |
| 6. 40* Illiterate Colombians aged 55–60 for Animals and Fruit (average per category): 11.2 (*SD* not given) | | | |
| 7. 40* Professional Colombians aged 55–60 for Animals and Fruit (average per category): 17.8 (*SD* not given) | | | |

**N* per cell not given; prorated for 200 subjects across 5 cells.

and Colombian illiterates score about 11 items per category, which is some 40% less than the Colombian professionals and 50% lower than the South African professionals.

* * *

How these differences arise, what they mean, and how they can be bridged are among the most intriguing and difficult questions in psychology: The route neuropsychology must take in beginning to answer them is sketched in the following chapters.

*Part* II

# THEORETICAL FOUNDATIONS OF CROSS-CULTURAL ASSESSMENT

*Chapter* 3

# Radical Environmentalism: Vygotsky, Luria, and the Historical Determination of Consciousness

The most attractive response—politically and intellectually—to the problems raised by test score variations across countries and cultures is radical environmentalism, which holds that culture makes mind. Proponents of this view, which evolutionary psychologists now deride as the "standard social sciences model" (Cosmides, Tooby, & Barkow, 1992), hold that culture is a semiotic system, that is a system of meanings extracted from signs:

> Believing, with Max Weber, that man is an animal suspended in webs of significance he himself has spun, I take culture to be those webs, and the analysis of it to be therefore not an experimental science in search of a law but an interpretative one in search of meaning. (Geertz, 1971, p. 5)

It is impossible to grasp the problems of cross-cultural psychological assessment, or to make sense of the furious debate about the meaning of IQ score differences between groups, without first considering whether culture makes mind, and, if it does, how this prodigious feat is achieved.

It is appropriate (and also neuropsychologically informative) to explore these difficult issues of mind, culture, and the ways in which they reproduce one another through the life work of the great Russian neuropsychologist, Alexander Romanovich Luria (1902–1977), and a replication of his cross-cultural work in rural South Africa in the 1980s. Luria's prodigious output of books, monographs, and papers laid the foundations for modern clinical neuropsychology (1963, 1970, 1973, 1975a, 1975b, 1976a, 1976b, 1976c; see also Scheerer & Elliger, 1980, for a bibliography of Luria's foreign language papers).

Luria was born in 1902 in the city of Kazan, an old university town on the Volga River, 600 miles east of Moscow. His father was a doctor, fluent in German (which Luria also acquired at an early age), and an accomplished intellectual, but was barred from a post at the University of Kazan because he was Jewish: The name *Luria* is old and distinguished in Jewish history. The Russian Revolution of 1917 had a liberating effect on the Luria family as it did throughout Russia. Quota restrictions were dropped, and the universities were opened to all. Soon after, the father took a post at the Kazan Medical School, and later established an independent institute for advanced medical studies in Kazan. Luria enrolled as a student, and, while engaged in the traditional university curriculum (including courses on Roman law, jurisprudence, and the like) his interest in psychology developed. He translated Brentano's *Theory of Human Drives* into Russian, and plunged into a study of Freud's *Interpretation of Dreams*, as well as early works by Adler and Jung. Luria established the "Kazan Psychoanalytic Association," and wrote to Freud to announce this event. He was delighted to receive a note from Freud expressing gratification that a psychoanalytic circle had been founded in such a remote eastern town.

After graduation from Kazan University in 1921, Luria's primary ambition was to become a psychologist, but he was under pressure from his father to study medicine. He did both: While doing the preliminary 2 years of medical studies, he also worked at the Kazan Psychiatric Hospital and continued reading psychology. In 1923, Luria was invited to join the staff of the Moscow Institute of Psychology, where his first research was on motor reactions to "neutral" and "critical" stimulus words, through which he examined "hidden complexes." In 1924, Luria met Lev Vygotsky (1896–1934), the most brilliant psychologist of his generation, who soon after joined the staff of the Moscow Institute, and became a formative influence in Luria's development as a psychologist. Luria (1979) wrote about Vygotsky as follows:

> It is no exaggeration to say that Vygotsky was a genius. Through more than five decades in science I never again met a person who even approached his clearness of mind, his ability to lay bare the essential structure of complex problems, his breadth of knowledge in many fields, and his ability to foresee the future development of his science. (p. 38)

## VYGOTSKY'S CULTURAL MATERIALISM

Vygotsky's work has become the theoretical base for those who see human development as a continuing process of change and growth rather than the achievement of a predetermined plateau (chap. 9). Vygotsky in turn had been profoundly influenced by Marx: Foucault (1980) reminded us that the power of Marxist thought is in its massive assimilation of historical processes rather than its economic theory. In essence, Marx took the view that thought is a product of the material conditions in which it develops. By acting on the external world through his labor, wrote Marx, man "at the same time changes his own nature. He develops his slumbering powers and compels them to act in obedience to his sway" (from Vol. 1 of *Capital*, 1867, cited in McLennan, 1981, p. 174). Labor is in turn carried out by the use of tools, which are a uniquely human aspect of the labor process: It is by study not of the tools themselves, "but how they are made, and by what instruments, that enables us to distinguish different economic epochs" (in McLennan, 1981, p. 176). Tools are internalized as ideas:

> A spider conducts operations that resemble those of a weaver, and a bee puts to shame an architect in the construction of her cells. But what distinguishes the worst architect from the best of bees is this, that the architect raises his structure in imagination before he erects it in reality. (p. 176)

The idea in the architect's mind is determined by the material characteristics of the culture in which he lives:

> In acquiring new productive forces, men change their mode of production, and in changing their mode of production, . . . they change all their social relations. The windmill gives you society with the feudal lord; the steam-mill society with the industrial capitalist. The same men who establish social relations in conformity with their material productivity also produce principles, ideas, and categories conforming to their social relations. (from *The Poverty of Philosophy*, 1847, cited in McLennan, 1981, p. 40)

This is a revolutionary idea, that the principal contents of cognition, its concepts and values, are the products of the material circumstances of life.

These ideas reemerge in psychologically elaborated form in Vygotsky's work. Kozulin (1984) commented that a psychological phenomenon could in Vygotsky's view "be comprehended only through a study of its origin and history" (p. 105). This study is necessarily of ontogenesis, that is, of individual development and especially the individuals' acquisition of psychological tools fashioned by the society in which they live. Vygotsky gave a zoological example: The whale described by appearance is a fish. This is a phenotypic analysis. By developmental study—that is, examination of its causal dynamic basis—this fish is shown to be "closer to a cow or a deer than a pike or a shark" (Vygotsky, 1988, p. 62). Preeminent among these psychological tools fashioned by society is language:

> The dialectical approach, while admitting the influence of nature on man, asserts that man, in turn, affects nature and creates through his changes in nature new natural conditions for his existence. . . . Human behavior comes

> to have that "transforming reaction on nature" which Engels attributed to tools. (Vygotsky, 1988)

In a much-quoted passage from his unpublished notebooks, Vygotsky (1988) wrote:

> The study of mind must be approached only after one has learned the whole of Marx's method. The whole of *Capital* is written according to the following method: Marx analyses a single living "cell" of capitalist society—for example the nature of value. Within this cell he discovers the structure of the entire system and all of its economic institutions. He says that to a layman this analysis may seem a murky tangle of tiny details. Indeed, there may be tiny details, but they are exactly those which are essential to "micro-anatomy." Anyone who could discover what a "psychological" cell is—the mechanism producing even a single response—will thereby find the key to psychology as a whole. (p. 8)

By this process, according to Vygotsky, in a wonderfully evocative phrase, people can create their own capital, playing here on the idea of intellectual capital in the sense of a valuable store of ideas and methods, and at the same time alluding to Marx's great work (there is a brilliant exploration of these themes in an essay by the Black South African psychologist, N. C. Manganyi, 1991, chap. 8).

It must in passing be noted that powerful demands for the qualitative analysis of test performance arise from Vygotsky's work. Giving a score is phenotypic, which makes of a whale a fish; describing its causal dynamic, in which the essence of the whale is contained, makes of the whale what it truly is. Qualitative analysis is thus ontogenetic in a Vygotskian sense. Again, Marx is helpful, remarking that "if the essence of artifacts coincided with the form of their manifestations, then every science would be superfluous" (cited without a source in Vygotsky, 1988, p. 63). Today, close to a century later, this observation is equally true of psychometrics: The score (i.e., the outer appearance) is not the essence. The essence is ontogenetic, namely, the underlying construct.

## PSYCHOLOGICAL PROCESSES AS HISTORIC

Luria articulated Vygotsky's work with his own in two seminal papers. The first began as follows:

> Forty years ago, in the mid-twenties, a young Soviet psychologist—he was only 30 years old—L. S. Vygotsky began his visits to a neurological hospital, in the beginning as an observer and after that as an independent investigator. . . . He did not come to apply already known tests for proving diagnosis of brain injuries. His task was incomparably wider: In the analysis of local brain

> damage he saw one of the basic ways of the analysis of the most important structure of psychological processes and a possible approach to the material substrata of complex psychic activity. (Luria, 1965, p. 387)

In a paper he completed 6 weeks before his death in 1934, Vygotsky wrote: "It seems to me that the problem with localization, as a general rule, includes all that is connected with the study of higher psychological functions and with the study of their disintegration" (in Luria, 1965, p. 387). But it was not the mindless locationism of faculty psychology or of the cortical map-makers that Vygotsky had in mind. In the *Development of Higher Psychical Functions* (1960), published in abbreviated form as "Problems of Method" in *Mind in Society* (1988), Vygotsky argued that the "higher psychological functions" ought to have their own origin, which was located not "in the hidden properties of nervous tissue [but] outside the organism of the individual person, in objectively existing social history which is independent of the individual" (Luria, 1965, p. 388). These functions, such as imagination, thinking, emotion, and voluntary activity, were not isolated functions but rather complex functional systems "formed in history and changing in the process of ontogenetic development" (p. 389). For humans, a new principle of localization is required, akin to the way a person ties a knot in a handkerchief in order to remind himself to carry out a particular action. The knot exists outside the brain, but is used to tell the brain, by the mediation of this sign, what action to take:

> Social history ties those knots which form definite cortical zones in new relations with each other. . . . The human cortex becomes, thanks to this principle, the organ of civilization, containing in itself unlimited possibilities and not demanding creation of new morphological apparatus every time a demand is created in history for a new function. (p. 391)

In this light, neither the narrow localization of faculties, or conceptualization of the brain as a single mass of homogeneous tissue is appropriate, but rather "a system of highly differentiated zones of the cortex working together, accomplishing new tasks by means of new inter-areal relations" (p. 391). The importance of these formulations for Luria's own work is immediately apparent, especially his still-uncontested hierarchical arrangement of cerebral function in terms of three integrated systems, and the related concept of syndrome analysis (see, especially, Luria, 1965, 1973).

In a second major elaboration of Vygotsky's views, "Towards the Problem of the Historical Nature of Psychological Processes" (1971), Luria examined the implications of the way in which the higher processes that make the brain the organ of civilization are created, which, in Vygotsky's formulation, is social in origin, mediated in structure, and voluntarily directed (p. 261). This, in turn, suggests that cortical structures will be modified by individual maturation and, necessarily, by social structures.

The rapid social change following the Bolshevik revolution of 1917 created a huge natural laboratory in the Soviet Union for verification of the Vygotskian hypothesis: Just as the acquisition of language changes the child's environment and thus the structure of consciousness, so would the processes of modernization, and especially schooling, change traditional ways of thinking.

## THE UZBEK EXPEDITION

In spring 1931, Luria and a group of colleagues from the Institute of Psychology in Moscow set off for the Soviet Republic of Uzbekistan in central Asia, deep in the Muslim Near East, with the Caspian Sea to the west, Afghanistan to the south, and the Chinese border to the south and east. Collectivization propelled the villages and nomad camps of these areas from feudalism to a 20th-century organization of production in the space of a few months or years. This process of accelerated social change was then and continues now, to a greater or lesser extent, in all the developing countries; Luria's remark that the "extraordinarily deep and rapid restructuring of historical forms" in the Soviet Republics "had never occurred before . . . in exactly the form experienced in the USSR" (1971, p. 265) evokes some irony.

The purpose of Luria's field study, which like all his work was meticulously planned and executed (but unpublished for 40 years), was to determine whether different groups of peasants, at different levels of modernization, performed simple intellectual tasks in different ways: Vygotsky's theory of cortical development by the "tying of knots" through the mediation of social experience predicted that the tasks would indeed be performed differently. The five subject groups identified by Luria were Ichkari women living in remote villages who were illiterate and not involved in modern social activities; peasants in remote villages with "an individualistic economy"; women who had attended short teaching courses but remained semiliterate; office holders in the collective system who were barely literate but engaged in the planning of production, organizing labor, and other activities that exposed them to new methods of problem solving; and women students who had been admitted to a teachers' school but still had fairly low qualifications.

The tasks were embedded in a learning situation in which the investigators showed the subjects new ways of solving problems to see how these would be incorporated into the actual tasks. Luria (1976a) distinguished between two modes of generalization. One is *graphic recall* based on memories of how certain objects in the individual's personal experience relate to each other in day-to-day activities: For example, a triangle, a trapezoid, and a circle might be grouped together because these remind the individual of a cooking pot (trapezoid) on a hearth (circle) under a tree (triangle). The other generalization mode is abstract, developed under the influence of formal schooling and based on isolating common attributes so that each object is assigned to an abstract category (e.g., "utensils" or "implements")—hence, the term *categorical relationships*, which are distinct from graphic or concrete relationships.

The unsophisticated[1] subjects were unable to form categories, but instead grouped objects in terms of their real life relations: For example, given a set of geometric shapes, they named them by their resemblance to everyday objects—a circle was called a plate, a triangle a kettle stand, a broken circle a bracelet, and so on—rather than grouping them by their form. Luria suggested the metaphor of a family for this kind of concrete classification, in which each member participates on an individual basis. On the other hand, subjects with some education and experience of modern methods readily made the transition to categorical relations or, in Luria's term, taxonomic thinking.

A second type of categorization task allowed for the flexible use of classificatory schemes. Subjects were presented with four items (e.g., drawings of a hammer, a saw, a log, and an axe; or a tumbler, a saucepan, spectacles, and a bottle) and asked "Which of these does not belong?" Given the first of these sets, an illiterate 30-year-old peasant said they were all alike, and all belonged: "See, if you're going to saw, you need a log, and if you have to split something you need a hatchet. So they're all needed here." Told that someone else had picked the hammer, saw, and axe, and omitted the log, the subject responded, "Probably, he's got a lot of firewood, but if we'll be left without firewood we won't be able to do anything. Even if we have tools, we still need wood." On the second set, he responded that the spectacles did not belong, but "Then again, they also fit in. If a person doesn't see too good, he has to put them on to eat dinner." Told of someone else who had omitted the saucepan, he responded, "Probably that kind of thinking runs in his blood." Luria noted that this kind of situational thinking (e.g., that one needs the tumbler, saucepan and bottle in the kitchen, and the spectacles to see better what one is cooking) was resistant to change, and there was consistent rejection in the unsophisticated group of the theoretical task in favor of the practical one: "Only classification based on practical experience struck them as proper or important" (Luria, 1979, pp. 69–71).

Classifications of groups of objects were based on practical experience and concrete operations, such as building a hut or damming a stream;

---

[1]A word derived from the Greek, meaning "uncorrupted by worldly wisdom," and strictly accurate here. However, as noted later in this chapter, Luria's belief that sophistication was qualitatively better than "primitivism" gave rise to a furor he had not anticipated.

syllogistic thinking was entirely absent. For example, subjects were presented with a major and minor premise as follows (in Luria, 1979):

> In the far north, where there is snow, all bears are white.
> Novia Zemlya is in the far north.
> What color are the bears in Novia Zemlya?

Subjects referred exclusively to their own personal experience did not accept the premises as universals, but as particular statements to be judged in terms of experience, and failed to construct logical links between the premises. They might respond, "I've never been in the north and never seen bears," or "There are different kinds of bears. If one is born red, he will stay that way" (Luria, 1979, pp. 77–78). With the lucidity that is his remarkable gift, Luria (1979) concluded that "the processes of abstraction and generalization are not invariant at all stages of socioeconomic and cultural development. Rather, such processes are themselves products of the cultural environment" (p. 74).[2]

## REPLICATION OF THE UZBEK STUDY IN KWA ZULU-NATAL

In 1984, Andrew Gilbert, a South African psychologist who was then director of the Institute for Social Research at the University of Zululand, undertook a replication[3] of Luria's Uzbek study in rural Kwa Zulu in South Africa's Natal province; to protect his subjects, he has not named the village.

Gilbert framed his study within an ideational definition of culture, which is seen not as complexes of concrete behavior patterns "but as a set of control mechanisms—plans, recipes, rules, institutions—for the governing of behavior" (Geertz, 1971, p. 44). Culture thus provides a link "between what men are intrinsically capable of becoming and what they actually, one by one, in fact become" (p. 52). Gilbert integrated this semiotic understanding of culture with the process of social change, and thus with

---

[2]For those wanting more detail of the Uzbek expedition, Luria's most extensive account is in his book, *Cognitive Development* (1976a), in turn based on a much fuller Russian version published in 1974. Attractive brief accounts of the expedition appear in Luria's 1979 autobiography and the 1971 monograph.

[3]Gilbert's theoretical structure and field study are more fully described in his doctoral dissertation (1986), which has not had journal or book publication, and will repay reading by those interested in the cognitive aspects of modernization, or working in the area of psychological assessment in developing country settings. A related examination of cognitive change under the impact of modernization is Berry's (1988) study of the Cree people of James Bay in northern Quebec in the decade of the 1970s, when a hydroelectric plant was built in their territory.

process rather than state, with becoming rather than being: The consequence is that culture cannot be reified as an immutable essence. Change is possible because men and women are not only the products of their culture, but also produce it, and (as shown in chap. 9) the neuropsychologist can play a small role in this process by becoming a "cultural guide" (Gilbert, 1986, pp. 161–162; see also Gilbert, 1989).

Gilbert selected five groups of five subjects, all born and raised in the study area, and corresponding to the five types identified by Luria 50 years earlier in Uzbek: The *poor* were subsistence farmers or casual laborers who had little formal education or city experience; *farmers* had no formal training in agriculture, but produced at a higher level than subsistence farmers; *entrepreneurs* were shop owners and businesspeople; *community workers* had no formal qualifications but were active in community-based projects; *professionals* were teachers and nurses who had studied away from home and returned to the area as carriers of modernity. Regarding education, the five poor people had an average of 1 year's schooling (three of them had none at all), the farmers had just over 4 years, the entrepreneurs nearly 7 years, the community workers 8 years, and the professionals nearly 12 years of formal schooling.

Asked to group geometric forms together, the poor subjects, as in Luria's study, "consistently classified the figures in a concrete object-oriented manner" (Gilbert, 1986, p. 201), despite extensive prompting. Shown a square, a circle, and a triangle, one of this group, for example, said, "This is a plot of ground (square), this is the circle that is drawn when you start building a hut, and this (triangle) is the roof." The professionals, in contrast to Luria's findings, did not as a matter of course use abstract categories, but on occasions concrete or graphical criteria and at other times geometric categories. In contrast to the poor, whose thinking was rigid and who refused to consider alternative classification methods, the professional group showed a consistent habitual response but, with prompting, moved flexibly to other methods of classification. The entrepreneurs showed similar flexibility, at times sorting in object-oriented terms, at times by graphic similarity, and sometimes by geometric category. The farmers, although they had less education and had spent far less time in cities than the entrepreneurs, showed a remarkably similar grouping style. One, with no education at all, flexibly used both geometric and graphic categories, showing a reflective and flexible style of thought.

On the categorization task, Gilbert's less educated subjects had difficulties that paralleled those Luria found in Uzbek. For example, one participant said that the log belonged equally well with the three other items because "One has to make a fire with this wood. Should it be taken out?" Another reasoned the same way, acknowledged on prompting that three of the items were metal, but at once reverted to a functional classification that included the wood because it was needed to make a fire. The professionals were sometimes able to think abstractly, but not habitually. Thus, one argued that the axe could be used on the log if you were building a house, but that the log could be excluded on the grounds that the other three items were made of iron, adding on reflection, "The handle of the axe and the hammer are made from a log," reverting to situational thought despite having used a category.

The community workers closely resembled the professionals in their classification style. The farmers differed, because although they preferred situational groupings, they were, unlike the poor, prepared to agree to abstract categorizations without dismissing these as incorrect: This is surprising, according to Gilbert, because the farmers had only marginally more education than the illiterate group.[4]

The entrepreneurs were all able to use abstract thought, although, paradoxically, the two best educated persons in this group veered between abstract and situational thinking while the other three used categories spontaneously and consistently. An example of the former style is the participant who said the saw and log go together because you cannot saw without a log, then, on prompting, added the axe to this group because they work together and, finally, said three of the items belonged together because all were made of steel. Thus, alternative groupings are supplied, some concrete-situational and the other categorical-abstract. Similarly, this subject excluded spectacles from the second group on the ground that "you can't put them in boiling water . . . they are specially designed for the eyes," but failed to identify the unifying category of "utensils" that would exclude the spectacles.

## CULTURAL DIFFERENCE AND THE SUPPRESSION OF SCIENCE

In 1933, Luria published a preliminary paper in the journal *Science* setting out the goals of the Uzbek expedition, but he did not publish his findings. Soviet psychology was under political scrutiny, and Luria's initial reports had attracted vitriolic criticism. In her unpublished memoir of her father, L. Luria recorded her father as saying, "I was accused of all mortal sins right down to racism, and I was forced to leave the Institute Of Psychology" (cited in Knox's introduction to Vygotsky and Luria, 1993, p. 15). An

[4]Gilbert's use of education as an index of cognitive development is open to the criticism made by Rogan and MacDonald (1983; see chap. 5) of the terms "schooled-unschooled" as too broadly defined to constitute a useful variable. They wrote that the quality of education "differs enormously from one area to another" (p. 310); rather than asking how much schooling an individual has had, one needs to know its content, and whether aspects that would contribute to cognitive development were present (see chap. 5).

amusing sidelight on the charged political atmosphere of the time—although it would have been anything but amusing to Luria—was that KGB agents were waiting on the platform of Moscow station for Luria on his return. This was because one of the first experiments the research team carried out on arrival in Uzbekistan related to classical visual illusions in order to demonstrate that Gestalt principles were the consequence not of brain structure, but of culturally transmitted modes of perception. Luria was so excited by the preliminary results that he telegraphed Vygotsky to say, "The Uzbekis have no illusions!" (Luria, 1979, p. 213). The KGB, intercepting the telegram, read it to mean that the Uzbeks had no illusions about Soviet authority in that area! (Knox, 1993, p. 14).

Kozulin (1984) described the reception given to the Uzbek work:

> Luria could hardly have imagined the bitter fate that awaited his studies. The Vygotsky–Luria theory of cultural development was already under fire because of its apparent resemblance to the "bourgeois speculations" of Emile Durkheim. Critics hastily accused Luria of insulting the national minorities of Soviet Asia whom he had ostensibly depicted as an inferior race unable to behave reasonably. The results of the expedition were refused publication, and the very theme of cultural development was forbidden for the next 40 years. Only in 1974 did Luria manage to publish his materials and thus to state the problem once again. (p. 110)

After Vygotsky's death in 1934, and dismayed at the reception of his work by Russian academics, Luria returned to his medical studies: Speculation is that he found these a safe haven from the stormy ideological seas on which psychology was adrift. In 1936, he enrolled as a full-time student at the First Moscow Medical School, where, because of his extensive previous coursework in medicine, he was able to graduate the following year. In 1938, he began a 2-year internship in neurology at N. N. Bourbenko's Neurosurgical Institute, which he described as "the most fruitful of my life" (p. 131). In this period, his serious work in neuropsychology began.

Without the encouragement of his American biographer, Michael Cole, Luria might never have published the Uzbek data at all. An initial theoretical paper on the historical nature of psychological processes appeared in the *International Journal of Psychology* in 1971, followed by the fuller 1974 account, published in English translation in 1976.

It is important to understand what it was about the Uzbek expedition, conducted in the heady early years of a great social revolution, that proved unpalatable to Luria's critics. L. Luria (in Knox, 1993) cited a 1934 review of the Uzbek expedition that appeared in the journal *The Book and the Proletarian Revolution*:

> Instead of showing the process of development and the cultural growth of the workers in Uzbekistan, they search for justifications for their "cultural

> psychological theory" and "find" identical forms of thought in the adult Uzbek woman and a 5-year-old child, dangling before us under the banner of science ideas, which are harmful to the cause of the national cultural construction of Uzbekistan. (p. 16)

Cole noted that Luria used the word *culture* in an evolutionary sense, which in turn implied that a qualitative difference would exist between "cultured" and "uncultured" persons: "Overall, his writings emphasized the 'improved' status of people following the advent of literacy and modern technology" (Cole, in Luria, 1979, p. 214). Careful reading of Luria's account of the expedition shows that he had indeed fallen into the pseudo-evolutionary trap that awaits incautious cross-cultural researchers at their first contact with people in a culture that differs radically from their own, who tend to attribute advancement to their own culture and primitivism to the other (as in Jung's account of his travels in east Africa; Nell, 1992). The Uzbek controversy for decades obscured the psychologically crucial issue Vygotsky and Luria attempted to address, namely, how mental processes are shaped by sociocultural forms.

Gilbert's work has had a less dramatic but parallel fate: It has sunk quietly into oblivion, ignored by the neuropsychologists and other students of cognitive processes who should be consulting it. Although we live in gentler times than Luria, and Gilbert's work has not been suppressed, its emphasis on culture and the impact of culture on cognition is out of tune with the universalism that psychology and neuropsychology have adopted as a defense against the excesses of "culturalism" (some of which were reviewed in chap. 1). At the close of the 20th century, the populations of developing countries have a critical awareness of the bitter fruits of colonial exploitation on which they and their children still choke. Angry memories of colonial arrogance are stirred by scientific or pseudo-scientific exploration of differences between ethnic or national groups.[5] However, between justice and the tragedy of opportunity denied (chap. 4) stands the need to recognize and deal with the reality demonstrated by the data in the previous chapter.

---

[5]See Nell (1999) for the South African legislature's response to these memories.

*Chapter* 4

# The Nature of Intelligence: The IQ Controversy in Cross-Cultural Perspective

As argued in the previous chapter, radical environmentalism is the most attractive response to the variability of test scores not only from one country to another, but also in the same country and within what appears to be the same culture. The purpose of this chapter is to consider the factors that determine the intelligence test scores of individuals and groups, and to reflect on whether these variations arise because of real differences in intelligence, or if they are artifacts of the tests themselves.

The chapter begins by considering the principle argument that is marshaled against environmentalist optimism. This is the nativist view, which holds that only some 40% (perhaps as little as 20%) of each individual's intelligence is shaped by the environment, with the remaining 60% to 80% genetically determined and therefore immutable. The often raucous debate between *environmentalists* and *nativists* (Herrnstein & Murray, 1994) has become known as the *IQ controversy*, and it is reviewed here not in its own right, but because it brings to a sharp focus the issues that emerge from the long history of psychological assessment in culturally different settings.

The IQ controversy thus serves as an introductory frame for this long and quite complex review. Thereafter, I have organized the discussion around the two major current paradigms of cognitive assessment: the *psychometric* and the *information-processing* paradigms. Two other measurement methods, both subcategories of the information-processing paradigm, are also important to this discussion: the *piagetian* paradigm (used here without a capital letter to indicate a class of intellectual operation derived from Jean Piaget's genetic epistemology), and *potential* assessment, which has since the 1920s seemed to offer the best alternative to the psychometric paradigm. This latter is a topic in its own right and is reviewed in its historical context in chapter 9.

## ENVIRONMENTALISM TRIUMPHANT, NATIVISM RESURGENT

As noted in chapter 1, IQ measurement had its beginnings in an era that equably accepted many racist beliefs. The paradigm shift that revealed the full ugliness of racist beliefs was set in motion by Nazi doctrines directed at extermination of "inferior" races (i.e., Jews, gypsies, and Negroes). By 1945, as the concentration camps were liberated and the horror of the Nazi genocide dawned on the outside world, the tide had turned, and customary or ideological racism, at least in scientific circles, was treated with the obloquy it deserved. Moreover, anything that smacked of genetic determinism was perceived as an odious throwback to racism.

For decades, the environmentalist view has therefore held sway, and mainstream psychology has paid only sporadic attention to the nativists. Into this new intellectual climate—shaped by memories of Nazi racism, by the triumphs of the civil rights movement, and by the growing success of the colonial struggle for liberation, a 1969 paper by Arthur Jensen, "How Much Can We Boost IQ and Scholastic Achievement?" burst like a bombshell. In the opening sentence, Jensen gave his view of the effectiveness of the compensatory education programs mandated by the U.S. government's Head Start Project: "Compensatory education has been tried and it apparently has failed." The reason for the failure, he argued, is that genetic factors are more important than the environment in determining intelligence: Environment accounts for about 20% of the mean IQ scores of a population, and heredity is responsible for the other 80%. The weightiest evidence for this 1969 conclusion are the studies of identical twins reared apart.[1] Because, as Jensen put it, "if their environments are uncorrelated, all they have in common are their genes" (p. 51). Cyril Burt, discredited after his death by allegations of forgery, found the IQ correlation between 53 such pairs of twins to be .77.[2] Heredity also determines

[1]The methodological problem with these studies, as H. F. Taylor (1980) and others pointed out, is that when twins are separated (following the death of both parents, e.g.), they almost always remain within the same family so that the environmental differences are not very large. If the environments differed dramatically, with one twin going to the home of Harvard professors and the other to an inner-city single mother on welfare (or to subsistence farmers in a developing country), the 80/20 heredity/environment loading the identical-twins-reared-apart studies have found might be very different.

[2]Herrnstein and Murray (1994) cited the Minnesota twin study (Bouchard, Lykken, McGue, Segal, & Tellegen, 1990) as yielding a correlation of .78, almost identical to Burt's disputed finding.

race, continued Jensen—and then comes the hard part: "On the average, Negroes test about 1 standard deviation (15 IQ points) below the average of the white population in IQ, and this finding is fairly uniform across . . . 81 tests of intellectual ability" (1969, p. 81). In *Bias in Mental Testing* (1980), Jensen argued his position on both these themes in greater detail: that intelligence is largely heritable, and that standardized psychometric tests were free of bias against minority groups in the United States.

The problem persists 25 years later. The 1994 publication of *The Bell Curve* by Herrnstein and Murray, the most controversial book in recent psychology, again brought the clash between nativists and environmentalists to the public domain, with literally hundreds of articles in the popular and scientific press attacking the nativist position.[3]

However, before attacking *The Bell Curve*, it is first necessary to understand the arguments relevant to the psychological assessment of culturally different people. Herrnstein and Murray (1994) wrote that by the 1980s, the state of received wisdom about IQ testing was that

> intelligence is a bankrupt concept. Whatever it might mean—and nobody really knows even how to define it—intelligence is so ephemeral that no-one can measure it accurately. IQ tests are of course culturally biased, and so are all other "aptitude" tests, such as the SAT [Scholastic Aptitude Test]. . . . The tests are nearly useless as tools . . . [and] do not predict anything except success in school. . . . All that tests really accomplish is to label youngsters . . . creating a self-fulfilling prophecy that injures the socioeconomically disadvantaged in general and blacks in particular. (pp. 12–13)

*The Bell Curve* argued on the contrary that IQ and what it measures are real. The authors cited data showing that IQ scores vary substantially across ethnic and cultural groups: Chinese living in Hong Kong have a mean IQ of about 110, Jews living in the United States have a mean of between 108 and 115, and Latinos living in the United States have a mean of about 85. For African Americans, they reported that even when rigorous selection criteria are applied to eliminate methodologically suspect studies, the Black–White difference of just over 1 standard deviation holds up; the National Longitudinal Study of Youth (NLSY), which administered an apparently bias-free IQ test called the Armed Forces Qualification Test to 6,502 Whites and 3,022 Blacks, found a Black–White difference of 1.21 standard deviations.

---

[3]A curious aspect of this polemic is that it has been focused on ethnic differences in cognitive ability. But the book's central thesis is that American society is today stratified not by social class, but by intelligence—a proposition that has significant policy implications, but has been largely ignored in the brouhaha about the politically much more sensitive issue of race and IQ.

### Heredity and the Social Construction of Hope

These data have profound human welfare implications. I can only begin to imagine the hurt and bewilderment I would have experienced had I been a member of one of these groups. As a child I would have been placed in the D-stream in my school regardless of my test scores (a practice called "racial tracking"), assigned to bored teachers who expected me to be dumb and treated me accordingly, and then, as I matured, forced to carry this burden of customary expectations that made it near impossible for me to get into a good college.

At the same time, the claim that IQ is immutable strikes at a profoundly held Western belief: the notion of human progress and perfectability. The Protestant Ethic and the triumph of free market capitalism rest on the conviction that each individual has limitless potential: Bellhops become hotel magnates, and a man born in a log cabin becomes president of the United States. The reach of this limitlessness is set by effort, not by inheritance. If on the other hand intelligence is fixed by race and by birth, then fundamental social beliefs are strained. Predestination rules, and the liberal-democratic dream lies in tatters. Even worse, an immutable barrier to the individual's progress threatens social stability by threatening the social construction of hope (Samuelson, 1994).

### The Problem of Bias

The obvious rejoinder to these conclusions is that IQ tests are biased against African Americans and Hispanics. One of many U.S. studies pointing to this conclusion is by Jane Mercer (1984), who compared the WISC–R performances of 627 Black, 617 Hispanic, and 669 White students, all native-born and all at California public schools. Her conclusion is "that the WISC–R discriminates systematically in favor of white students at the item level, the subtest level, and the scale level. . . . Inferences based on those test scores are racially and culturally discriminatory." With regard to other cultures, chapter 2 and the remainder of this chapter review many different kinds of evidence of internal bias.

Herrnstein and Murray (1994), on the contrary, concluded that "no one has found statistically reliable evidence of predictive bias against blacks"(pp. 281, 627). Moreover, "the cultural content of test items is not the cause of group differences in scores" (p. 282). These sweeping statements sit strangely with a paragraph from the last chapter of *The Bell Curve*:

> The gaping cultural gap between the habits of the underclass and the habits of the rest of society, far more impassable than a simple economic gap between poor and not poor, or the racial gap of black and white, will make

> it increasingly difficult for children who have grown up in the inner city to function in the larger society even when they want to. (Herrnstein & Murray, 1994, p. 524)

In a strictly technical sense, this acknowledgment of cultural difference can be reconciled with the denial of bias. This is because IQ tests predict the school and university achievements of Black Americans no less accurately than those of Whites. In other words, when *The Bell Curve* says that IQ tests are unbiased, it is not talking about bias in the tests, but about *external bias*, which is measured against the criteria of school and college success.

Of course, the criterion might be as biased as the test.[4] After all, the educational system and the test system are devised and implemented by a single intellectual elite that shares a wide range of work-related and cultural values, and the two systems are two sides of the same coin: Just as academic success can be predicted by IQ test results, so should it be possible to predict IQ by extrapolating from academic success. It is therefore tautologous to say that tests are unbiased when measured against a very similar kind of yardstick.

***The Pernicious Consequences of Inflating IQ Scores.*** The damage done to individuals whose IQ scores are artificially lowered by test bias is clear. But giving some people artificially higher scores also has a pernicious effect. Commenting on the Spanish version of the WAIS published in 1968 and still in use in the United States for the assessment of Spanish-speaking adults, Melendez (1994) noted that it inflates full scale IQ scores by about 20 points in comparison with the U.S. versions of this test. This has the effect of labeling impaired individuals as normal and thus depriving them of social benefits and services to which they would be entitled if their IQs were correctly computed. Melendez also noted that at the time this version was standardized,

> the concept that a large segment of the Puerto Rican population had significantly lower IQs than the comparative sample of the WAIS must have been both scientifically and politically unsettling. . . . However, it is both demeaning and patronizing toward Hispanics to use a test . . . which artificially boosts the IQ results. . . . The transformations caused by the [Spanish version of the WAIS] may be an egalitarian's dream, but they can also be a clinician's nightmare. (p. 392)

Similar difficulties have been noted (Nell, 1994) for the South African version of the Wechsler Adult Intelligence Scale (which is not in fact the

[4]Herrnstein and Murray acknowledged this possibility (pp. 285–86), but went on to dismiss it as a possible source of test bias.

WAIS but the Wechsler–Bellevue, an earlier and cruder version that predates the WAIS by 16 years and the WAIS–R by 42 years): The South African test inflates full scale IQ by close to a standard deviation, and also has the effect of depriving significantly impaired individuals of compensation.

The Hispanic WAIS and the South African Wechsler–Bellevue inflate IQ scores unintentionally. But setting norms at a spuriously low level can take on a sinister aspect, for example if an insurance carrier promotes unduly flattering norms in order to "prove" that individuals who have been significantly compromised by head injury are still normal!

***Broadening the Canvas.*** For an outsider like myself, the most curious aspect of *The Bell Curve* debate is its insularity. It is conducted as if the United States were the whole world.[5] What might happen to the IQ controversy if the canvas were broadened to take account of a nearly a century of cross-cultural psychological assessment? The following sections of this chapter—organized as noted previously around the psychometric and information-processing paradigms of psychological assessment—provide additional perspectives on culture and intelligence. In subsequent chapters, armed with this additional knowledge, we can reflect again on the issues of central concern to us: Why do test scores differ across cultures? How can practicing neuropsychologists deal with these variations? And, how can this variation be addressed so that the neuropsychological assessment of culturally different subjects produces not misleading inferences of deficit, but valid and clinically helpful findings?

The following detailed material on test performances in culturally different settings should be read throughout as an extended commentary on the IQ controversy, and as a marshaling of the evidence for the conclusions reached at the end of this chapter.

## THE PSYCHOMETRIC AND INFORMATION-PROCESSING PARADIGMS

A useful way of distinguishing between psychometrics and information processing is to see the psychometrists as classicists and the information-processing school as revisionists (Herrnstein & Murray, 1994). The *classicists* see intelligence as structure: They work within the tradition of Spearman, whose 1904 paper identified a general intelligence factor, *g*, that stands at the center of this structure. "Despite numerous theoretical attacks on Spear-

[5]The United States is in any case a poor laboratory for the study of the effects of culture on intelligence. Looking at the available evidence, Hunt (1995) observed that there is "surprisingly little evidence for influences of cultural experiences on intelligence" (p. 365). This is very likely because such evidence has not been seriously sought in a society that prides itself on universal adherence to core cultural values.

man's basic notion of a general factor, *g* has stood like a Rock of Gibraltar in psychometrics, defying any attempt to construct a test of complex problem solving which excludes it," wrote Jensen (1969, p. 9), who is the foremost of modern classicists.

The *revisionists*, on the other hand, emphasize process rather than structure. Their focus is on information processing, on what people do when acting intelligently. Jean Piaget was the first of the revisionists, and Robert Sternberg is the leading contemporary worker within this paradigm.

However, like all dichotomies, the distinction between the psychometric and the information-processing paradigms is initially helpful, but becomes forced and artificial if pushed too far. Ultimately, there is no dividing line. The literal meaning of psychometrics is the measurement of intellectual processes, and this by definition includes all forms of mental measurement. There is in fact an inexorable movement from psychometrics to information processing: Psychometric structuralists become information-processing revisionists as they seek explanations for their findings.

This interparadigmic movement is well illustrated by Jensen himself. In the concluding section of his conventionally structuralist 1969 paper, he accounted for test performance anomalies by proposing a two-level structure of intellect. *Associative ability* (Level 1) is tapped by tests such as digit span, serial order learning, the learning of paired associates, or free list recall. Lower class children (White, Black, or Hispanic) perform as well on these tests as middle-class children. However, *conceptual ability* (Level 2) requires self-initiated transformation of the input before the response is made. Raven's Progressive Matrices, with its high loading on *g*, is a good example.

Jensen went on to note that a slight variation in the test procedure can change a free recall task from Level 1 to Level 2. For example, when a 20-word list made up of five items from each of four categories—animals, furniture, clothing, and foods—was used,[6] the lower class children did no better than they did on the uncategorized lists, and the middle-class children did much better. The latter's mean scores were about a standard deviation above those of the lower class group, with far more clustering by category (Jensen, 1969, p. 113): In other words, the middle-class children are able to apply a conceptual transformation to the input material. Of course, this is no longer structuralism, but squarely within the information-processing paradigm.

## I. The Psychometric Paradigm

In a 1935 address to the American Association for the Advancement of Science, Florence Goodenough remarked that intelligence tests are not

[6]The California Verbal Learning Test (Delis, Kramer, Fridlund, & Kaplan, 1990) makes use of exactly this paradigm.

measuring devices, but sampling devices. To determine what mental activities subjects engage in when confronted with a cognitive task, she continued, an experiment rather than an intelligence test is required (cited in Laboratory for Comparative Human Cognition, 1982). Encapsulated here is a major difference between the psychometric and information-processing paradigms. Everyday life is multifactorial, and so are psychometric tests: They sample broad swatches of behavior, such as engaging in a drawing-room conversation (the Information Subtest in the Wechsler), or explaining why a particular course of action should be taken (Comprehension), assembling a jigsaw puzzle (Object Assembly), or working out in one's head if a shopkeeper's bargain is such a bargain after all (Arithmetic). These behavioral samples have turned out to be most closely related to school situations, and for this reason they have very high predictive validity for success in the educational system.

But precisely because these tests are broad spectrum, they tell us very little about the component process out of which each of these samples of behavior is constructed. For example, in order to do well at Object Assembly, one needs to have visuomotor coordination and manual dexterity, to construct a mental model of what one is about to build, and to categorize it by giving it a name (e.g., "a child," "a camel"), to develop an edge alignment or internal detail strategy (Kaplan, 1988), to make effective use of error feedback in order to scrap false starts and begin again, and to possess the ongoing drive and future orientation needed to persevere.

The score alone reveals nothing about each of these separate components. Neuropsychology's response has been twofold: to retain psychometric tests because of their proven diagnostic utility, but to resort increasingly to qualitative test interpretation; and to supplement psychometric tests by information-processing probes.

Two aspects of the psychometric paradigm are of special importance in the testing of culturally different subjects: education effects and practice effects. However, these influences are pervasive and also relate to the information-processing paradigm.

***Education and Urbanization.*** Formal schooling plays a major part in all test performance, and overwhelmingly so in psychometric tests. Kendall et al. (1988) specified the elements of classroom skill that contribute to test performance: practice in using a pencil; familiarity with the use of booklets; facility with letters, numbers, and other symbols; an appreciation of the importance of paying attention, obeying instructions, and sitting still as contributors to speed and accuracy of work; and, in general, an appropriate orientation to the examination situation: "No other cultural learning experience is as concentrated and as fundamental than that which is provided, systematically, through the formal education system" (p. 310).

Powerful education effects recur with unfailing regularity in the testing of non-test-wise subjects, demonstrated for example by Crawford-Nutt's (1977a, 1977b, 1977c) work on the Symmetry Completion Test, administered to 1,151 Black subjects whose educational level ranged from virtually none ($M = 1.6$ years, $SD = 2.2$) to university students with a mean of 14 years of education ($SD = 0.7$): Their scores varied in almost linear fashion with education. Similarly, among South African paint factory workers (Nell et al., 1993) and farm laborers (Nell, Kruger, Taylor, Myers, & London, 1995), years of education was the single largest moderator of test performance.

An interesting sidelight on education effects is that in the youngest age groups, where no education effects exist, scores in different cultures converge more than at a later age. Verster and Prinsloo (1988) reported no differences between 3- to 5-year-old English and Afrikaans speakers on the Junior South African Intelligence Scale; but from age 6 differences began emerging, which suggests that differences are smaller before the commencement of formal schooling. Similarly, Richter and Griesel (1988) compared 722 Black South African children from 2 months to 30 months old with the 1969 U.S. reference group on the Bayley Scales of Mental and Motor Development. The South African children significantly outperformed the Americans at the ages of 4, 5, 6, 8, 10, 12, and 15 months on the mental scale scores, and at 2, 3, 4, 5, 6, 8, and 10 months on the motor scale scores ($p < .01$ on all groups except 10 months, $p < .05$). In view of the long-standing claim that African infants are precocious in relation to Western children, it is worth noting that when the South African and U.S. samples were simultaneously compared with British and Baganda infants tested on the same tests, there were no significant differences between the four groups (Richter-Strydom & Griesel, 1984).

Formal education effects greatly overshadow the consequences of urbanization when both are controlled, although there is nonetheless an acculturation construct that traces a hierarchy from rural illiterate to urban illiterate to rural literate to urban literate (Kendall et al., 1988, p. 312)—much as Gilbert (1986), Berry (1988), and other rural researchers demonstrated.

Between them, these factors of formal education and urbanization override ethnicity as a contributor to test performance variance in culturally different settings, and are also more important than the traditional sources of variance found in age, sex, and socioeconomic status: The "cultural variable," in which education and urbanization are subsumed, therefore makes far and away the largest contribution to performance variance on psychometric tests. Until shown otherwise, it must be assumed that these large cultural differences also occur between different groups in Western societies.

***Practice Effects.*** M. A. Verster (1976) described how the test performances of a group of 1,200 Black miners were affected by exposure to the technologically sophisticated mine environment on the one hand, and repeated testing on the other. The test instrument was the Classification Test Battery developed by the National Institute for Personnel Research of the South African Council for Scientific and Industrial Research. One group of subjects was retested four times at 3-month intervals to maximize test practice; another was retested only once, 12 months after the initial test administration, in order to maximize the effects of exposure to the highly Westernized mine environment; intermediate groups were retested two and three times, respectively.

Exposure to the mine environment did not make a significant contribution to test performance, but repeated test exposure did. The greatest mean increase, on the order of one half of a standard deviation,[7] occurred between the first and second testing, with diminishing increments thereafter, of the order of a quarter to a fifth of a standard deviation on the third and fourth retesting. Score increments thus describe a classic learning acquisition curve (Verster, 1976, Figure 3), with rapid initial learning that approaches asymptote after the fourth retest (Kendall et al., 1988, p. 307). Test score improvements did not vary significantly with initial test performance, indicating that in this largely illiterate group, the brighter subjects benefited as much from repeated test exposure as the duller subjects.

In a reexamination of these data to examine the sources contributing to this strong learning curve, J. M. Verster and Muller (1985) considered five possible contributors. They found that test items, test format, and mental operations involved in the underlying construct all contributed significantly to improvements, but not to test procedure and environmental stimulation. Another finding of note was that the educational level of subjects also influenced the amount they gained from retest exposure, with the least educated and most educated groups showing the highest gains.

For neuropsychologists working with subjects who have had very little previous test exposure, the question that arises is whether "in any one test exposure, one is measuring [a fully developed ability, or] at a point on the acquisition curve for the ability being measured?" (Kendall et al., 1988, p. 308). This question was addressed by Crawford-Nutt (1976), who showed that modifying the instructional phase in the presentation of Raven's Progressive Matrices eliminated the often-reported "inferiority" of Black subjects on this test.

***"Retarded" Westerners.*** But the eight-decade-long psychometric test enterprise, from 1915 to the present, with the Black peoples of southern Africa raises a wider question: "We are still far from a scientifically defensible understanding of the manner in which people from different cultures process the same information" (Kendall et al., 1988, p. 328). Psychometric

[7]Equivalent to 7 or 8 scale points on an IQ test, for example scoring 96 at the first testing and 104 at the second.

tests in Africa and in other developing countries have focused on industrial selection. Indeed, a perennial frustration of knowledgeable cross-cultural psychologists has been that procedures based entirely on well-known local activities and skills would have had no predictive value for industrial applications. There has thus seemed to be no need to determine indigenous concepts of "smartness": Sarcastically, Kendall et al. (1988) asked how well Westerners would do at discerning rhythms and counterrhythms in African music, constructing arguments that cannot be logically refuted, or deriving secondary meaning from various forms of visual symbolism:

> Psychologists would do well to consider just how "simple" and "retarded" Westerners would appear to black people conducting imaginary investigations of "intelligence" using African-designed techniques of evaluation. . . . The surest solution to the "problem" would be to accelerate the process of Africanization of Westerners over successive generations of contact with an African culture (p. 328).

## II. The Information-Processing Paradigm

The strength of the information-processing paradigm is in the decomposition of complex behaviors into their constituent parts, and the separate investigation of each of these constituents. For such investigations, the computer is both a processing model, and also a device that allows for response quantification along otherwise inaccessible dimensions, such as reaction time, movement time, or variations in response latency. Because laboratory tasks can be simplified to a point where all intact individuals can do all the items (see "Speed and Power" later), it is these dimensions, rather than success or failure, that are the dependent variables. Such single factor, low ceiling laboratory tasks are needed in order to tease out the different variance components that enter into a score to determine its level.

Although a multitude of information-processing models of complex problem-solving behavior are available, one above all has exceptional power in elucidating the precise nature of the differences between test-wise and naive subjects, and in accounting for the large differences in test scores between countries and cultures: this is Robert Sternberg's triarchic theory of intelligence (1984, 1986). The triarchic theory is analyzed below in substantial detail because of the lucidity and theoretical elegance it brings to the understanding of test score variations across cultures.

### Linking Psychometrics and Information Processing: Sternberg's Triarchic Theory of Intelligence

The triarchic theory is appealing because it takes into account both universal features of human intelligence, identical from one culture to another, as well as fundamental variations in the nature of intelligence and the ways in which it is appropriate to measure it from one culture to another (Sternberg, 1988). Also, the theory avoids trivializing "intelligence" out of existence as a human adaptive capacity, as it is both in a pure information-processing approach and by those theories that equate it with cortical speed (see the comments by Eysenck and by Jensen in the issue of *The Behavioral and Brain Sciences* in which Sternberg, 1984, appears).

The three components of the triarchic theory allow the construction of a cognitive science framework within which neuropsychological tests can be deconstructed and interpreted. They are a *contextual subtheory* that relates intelligence to adaptive mental activity in the person's real-world environment; an *experiential subtheory* that is two-faceted, relating novelty and automatization to task performance; and a *componential subtheory* that specifies the mental mechanisms or component processes through which intelligent behavior is effected.

### 1. *The Contextual Subtheory*

This subtheory "addresses the question of which behaviours are intelligent for whom, and where these behaviours are intelligent" (Sternberg, 1984, p. 269). As Irvine (1969) pointed out in another context, counting to the base of 12 and 20 was an essential skill in an England that had pounds, shillings (20 to the pound), and pence (12 to the shilling) for its currency; today, after decimalization, the capacity to count to such bases remains as a cognitive competence, but cannot be defined as intelligent, or included in tests that measure contextual intelligence. Defining contextual intelligence against an external standard—"purposive adaptation to, shaping of, and selection of real-world environments relevant to one's life" (1984, p. 271)—not only accommodates the Marxist view that mind is shaped by environment, but achieves a larger purpose: It gives formal recognition to cultural relativism, making it possible, without theoretical strain, to predict for example that performance on a supposedly universal measure of ability to deduce rules, such as the Halstead Category Test, is anything but universal. In cultures in which these deductive skills are used to adapt, shape, and select environments, it can be assumed that people have cognitive competence in eduction, and that this competence is there to be measured; in cultures to which these criteria do not apply, low scores on the Category Test will either have a different meaning or no meaning at all.

### 2. *The Experiential Subtheory*

This two-faceted subtheory is concerned with skills at the interface between individuals and tasks: the ability to deal with novel kinds of task and novel situational demands on the one hand, and the ability to automatize information processing on the other. *Automatization* is a central concept in cognitive psychology. It means the ability to perform a complex task without thinking about it. Experienced surgeons can graft a vein while telling jokes to one another, and experienced drivers can dictate a business letter or listen with full comprehension to the morning news while at the same time shifting gears. Full conscious attention (what Sternberg called "the global processor") is given to the former task, and none to the automatized driving task. Novice drivers, on the contrary, must dedicate the global processor to driving, and have no attentional capacity left over for anything else.

Of the three subtheories, the experiential is the most important for psychologists working with subjects who are not test-wise. Readers of the material that follows will recognize in it a reformulation, in more rigorous terms, of the question raised by Kendall and his coauthors (1988) in the context of practice effects: whether the test score of a subject from a different culture reflects a fully developed ability, or "a point on the acquisition curve for the ability being measured" (Kendall et al., 1988, p. 308).

*2.1* ***Novelty.*** Novel tasks are not automatized: To use Sternberg's evocative term, they are "nonentrenched." In this sense, Part A of the Trail Making Test (Fig. 4.1), which requires a subject to connect circles numbered 1, 2, 3, and so on, calls on entrenched knowledge, whereas Trails B, which requires the subject to connect 1 to A, A to 2, 2 to B, B to 3, and so on, deliberately sets up a conflict between two highly entrenched kinds of information, the number series and the alphabet. Similarly, to reflect further on the neuropsychological armamentarium, the first two parts of the Stroop Color-Word Interference Test call on entrenched knowledge, that is the reading of single-syllable words and the naming of colors; Part 3, the interference task, requires the subject to say aloud the color of the ink in which a word is printed rather than the word itself: If the word RED is printed in green ink, the subject must say "green," which again sets up a deliberate conflict between an entrenched process (reading the word you see printed on the page) with a nonentrenched task (naming an ink color and not the word printed in that color).

The two-facet subtheory thus measures intelligence "precisely at those points where the relation between the individual and the task or situation is most rapidly changing" (1984, p. 276). The change arises because the individual automatizes every novel task as rapidly as possible, so that one is measuring learning on a steep acquisition curve. This curve will reach asymptote at different points for more and for less intelligent individuals so that the difference in information-processing speed early and later in the task, after asymptote has been approached or attained, will in fact be a measure of the time the individual needs to convert novelty to automa-

## Trail Making

### Part A

Sample

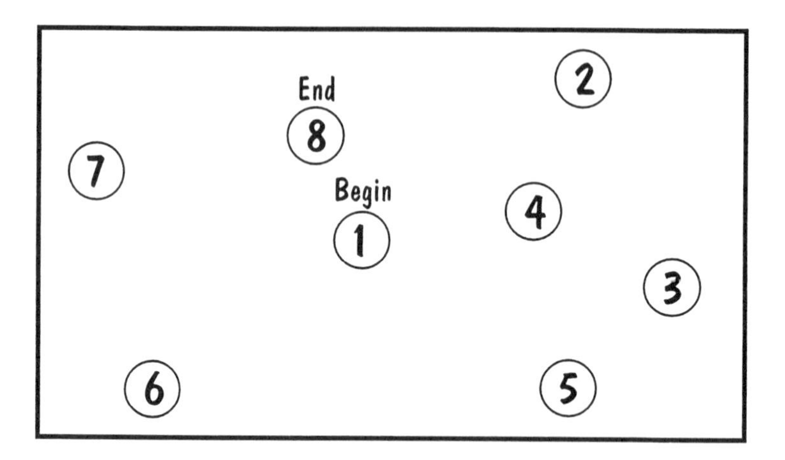


### Part B

Sample

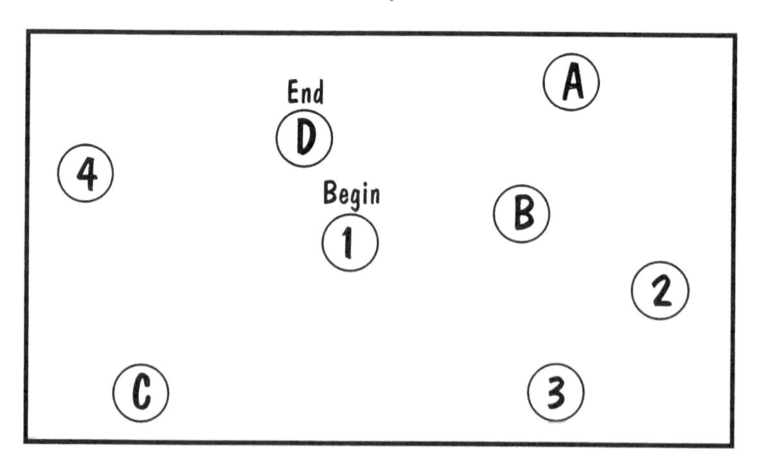


FIG. 4.1. The Trail Making Test.

tization. This in turn suggests that the measure of interest on a novel task, whether this is Part 3 of the Stroop, Trails B, or learning to change gears, will be the extent to which response latency decreases as a function of the number of items completed, or the number of gear changes practiced. Neuropsychologists, on the contrary, have hitherto concerned themselves only with elapsed time to task completion, or the number of responses in a stipulated time. Sternberg's theory suggests that a more interesting measure on Trails B would be to average response times from 1 to A, A to 2, 2 to B, and so on, in the first time epoch (this might be 10 or 15 seconds), and compare this average with average response time per item in the second epoch, third epoch, and so on. Time decreases from epoch to epoch would provide a numerical index of automatization, indicating how much automatization was taking place and its rate of increase or decrease.

Both tasks and situations can be novel. Novel or nonentrenched *tasks* require information processing outside of people's ordinary experience (thus, on a novelty–familiarity dimension, the Wechsler Verbal scale consists of familiar classroom-type tests, whereas the Performance scale items are novel, some more than others: Thus, Block Design is almost completely novel, and Object Assembly calls on jigsaw puzzle skills familiar to Western subjects). Sternberg noted that if a task is so novel that it is totally outside the individual's past experience (such as a syllogism for a Uzbeki peasant or calculus for a 5-year-old), then it has no assessment value.

*Situations* can also be novel or familiar, as with an examination taken in one's own classroom or in the hall of another school under a strange supervisor. Recall that the psychological test-taking situation calls on a range of skills familiar from the classroom situation, such as sitting still, paying attention, and using booklets (Kendall et al., 1988), which are unfamiliar to unschooled adults, thus adding a dimension of situational novelty to the high degree of task novelty.

A further distinction—that between the skills necessary to understand the task (task comprehension) and those needed to execute it (task solution)—can now be made within the area of task novelty. To measure *task comprehension,* Sternberg presented subjects with a "concept projection" task: An object could be presented either as a blue dot or a green dot, or described as one of four colors: *green, blue, grue,* or *bleen.* In the year 2000, the color might change, so that an object that is green now might appear blue in the year 2000: The name of its color in that year would be "grue"; similarly, what is blue now and green then would be called "bleen." The novelty demand of this task is to understand what is wanted (task comprehension) rather than to produce the novel words (task solution). (Having just read this section of Sternberg's paper several times in order to present the above description of the task, I can assure you that the task comprehension demand is indeed stringent!)

*Task solution* is assessed, for example, by a time–rate problem: If there are 3 windmills that each fill one tank in 2 days, how long will 12 windmills take to fill 3 tanks? Here, in contrast to the bleen–grue problem, task comprehension is easy for well-schooled subjects, but task solution is difficult because of the novel execution demands set by the problem. Sternberg called these insight problems, and by manipulating problem content, he and his colleagues were able to isolate three different kinds of insight subjects used to generate solutions: selective encoding, selective combination, and selective comparison.

*2.2 **Automatization.*** As already noted, familiar complex tasks such as reading or driving are completely automatized. However, novel tasks require fully conscious control by the global processor or central executive. This cognitive manager plans, monitors, and revises information processing, gets direct feedback from automatized nonexecutive processes, and makes available to them the individual's total knowledge base stored in long-term memory (Sternberg, 1984, pp. 277–278).

But the global processor is not needed for entrenched tasks, such as speaking one's own language or doing the jobs a busy executive carries out every day. For example, a sales executive dials a customer's number and greets the switchboard operator while at the same time filling out a factory work order on a familiar form: Here, the central executive calls two local production systems into operation, one for making the telephone call and the other for filling in an order form, both drawing on their own stores of locally applicable knowledge, while a third local system, talking, proceeds in parallel with both other sets of activities. Such automatized processing is of virtually unlimited capacity and does not require focused attention. But introducing a single element into this scenario that calls for the use of full conscious attention (e.g., the company buyer comes on the line to discuss the details of a new order) brings the other local processes to a complete halt: The notorious bottleneck in processing capacity caused by the use of full conscious attention (Kahneman, 1973) has come into play.

***Why the Wechsler Intelligence Scales Cannot Rule Out Brain Damage.*** The Australian neuropsychologist Kevin Walsh tells an anecdote to illustrate the consequences of prefrontal injuries for adaptive behavior. The story concerns a judge of the Australian High Court, to whom the excellent Wechsler-type IQ scores of a grossly hypofrontal traumatic brain injury survivor had been presented. The psychologist for the plaintiff then explained the nature of the man's adaptive deficit, to which the judge responded: "I see. You mean he has the intelligence, but he can't use it."

A troublesome shortcoming of Wechsler-type tests is that because they do not distinguish between entrenched and nonentrenched skills, they are blind to the adaptive syndrome produced by frontal injuries. Five of the six subtests on the Wechsler verbal scale (Information, Comprehension, Arithmetic, Similarities, and Vocabulary) draw heavily on old learning and classroom-type skills that are well entrenched and relatively more resistant to central nervous system insult than the novel and relatively unfamiliar skills required by the five subtests of the performance scale. In Sternberg's terms, because the Wechsler verbal scale stands on the automatization side of the interface between novelty and automatization, there can be startling contrasts between apparently preserved global intelligence on the verbal scale on the one hand, and grossly compromised life skills on the other.

*2.3 **Expertise.*** The disadvantage of the local production system is that it has available only a limited knowledge base that previous experience has packed into it. This limited knowledge base can be enormously expanded by experts in their field of expertise.[8] For example, a chess grandmaster has packed a huge amount of information into a local system and can use it in an automatized mode to handle 20 simultaneous chess games, and win them all. Part of the efficiency of such local systems derives from their nonhierarchical nature, integrating executive and process functions into a single system; if the local system is inadequate for a given condition, control is passed back to the attention-demanding global processor that decides how to handle the task. The benefit is that the new learning acquired during this exit to centrally controlled global processing is packed back into the local system (Sternberg, 1984, p. 278) so that there is no need to exit the local system next time round.

A useful model for the acquisition of expertise is thus created: It is "the successively greater assumption of information processing by local resources" (p. 278). Accordingly, novices are overwhelmed by new information when they enter an expert area, such as an upmarket hi-fi shop (or a neuropsychological case conference), and lose most of the new information as it is presented. Experts in the area have already stored a great deal of information in local systems, and have more global resources free—including selective attention—in order effortlessly to learn the context-bound new information, such as the wattage of the amplifier, the impedance of the speakers, and the oversampling ratio of the CD player; the novice, on the other hand, still puzzling out which of the black boxes is

[8]In arguing for the utility of mental speed as a principal contributor to intelligence, Jensen (1984) made a similar point: Because of the limited capacity of conscious attention, speed allows more operations per unit time to be executed without overloading the system, and before rapid decay has destroyed incoming traces. Thus, the most discriminating test items would be those that "threaten the information-processing system at the threshold of breakdown" (p. 295), which in turn could be predicted for more complex tasks by reaction time measures.

the amplifier and which the CD player, would have taken in no information about wattage or impedance.

*2.4 Automatization and Test-wiseness.* The more efficient the individual is in dealing with novelty, the more resources are available for automatized performance, noted Sternberg (1984). The trade-off between global and local processing takes place along an experiential continuum, and this has an important implication for the measurement of intelligence. For this purpose, two points of most interest on this continuum are the time when the task is first encountered, and the time on the experience acquisition curve when novelty wears off and automatization begins. For some subjects on some tasks—for example, for test-wise students taking a multiple-choice synonyms test—task comprehension is automatized. For others, task execution is automatized, as in a letter recognition or letter matching task. In both cases, some automatization will be acquired on the nonautomatized aspect while performing the task.

For ability measurement, Sternberg advocated selecting tasks that involve a blend of automatized and novel behaviors by presenting a task that is novel, but giving enough practice for performance to become "differentially automatized across subjects over the length of the practice period: Such a task will thereby measure both response to novelty and a degree of automatization, although at different times during the course of the testing" (Sternberg, 1984, p. 280). This formulation allows test performance to be related to cultural variables:

> Individuals who have been brought up in a test-taking culture are likely to have had much more experience with [both verbal tests and test of nonverbal reasoning] than individuals not brought up in such a culture. . . . Even if the processes of solution are the same, the degrees of novelty and automatization will be different, and hence the tests will not be measuring the same skills across populations. . . . Between-group comparisons may [thus] be defective and unfair. A fair comparison between groups would require comparable degrees of novelty and automatization in test items as well as comparable processes and strategies. (1984, p. 280)

For the neuropsychologist, the implications of this statement are dazzling. Even on apparently culture-fair tasks, such as Object Assembly or the WISC–R Mazes or the Austin Maze (Walsh, 1985, p. 236), Western subjects can use automatized task comprehension. Not so for subjects from other cultures. On the Austin Maze, for example, the arrays of switches and signal lights have analogues in such common objects as a puzzle toy or an instrument display panel. Automatized task comprehension will also be applied to the instructions for the test (in essence, "find a pathway through the array"), which is analogous to tracing through a maze, troubleshooting in an electrical system, and the like. This leaves global resources free for automatization of the initial phases of task execution, so that on each repetition of the maze, a larger number of correct moves has been moved into local memory.

The unsophisticated subject, on the other hand, can bring no automatic task comprehension to bear. The switch array on the board may be the first they have encountered. The instructions about finding a safe pathway across the buttons are obscure, even in amplified form, so that task comprehension requires full global processing capacity, leaving none to spare for automatization of task solution. For this hypothetical subject, then, in the first several trials, one is measuring not a visuomotor analogue of a supraspan learning task, but apparatus familiarization and instruction acquisition tasks. An unknown number of initial trials is thus used on tasks other than learning the pathway.

The experiential subtheory therefore specifies what a task or a situation must measure in order to assess intelligence: Tasks follow from the subtheory rather than the theory attempting a post hoc explanation of task demands. This in turn links intelligence to the real world rather than to tasks. Behavior is intelligent when it involves either adaptation to novelty, or automatization of performance, or both.

The issue of practice and test-wiseness that has been a perennial puzzle to ability assessment in the developing countries is also elegantly resolved. Practice facilitates both task comprehension and task solution, and is therefore an integral part of the task itself. All subjects must be given enough practice to allow for differential automatization of performance across subjects over time, so that the task will measure both response to novelty and degree of automatization at different times during the course of testing (Sternberg, 1984, p. 280). There are echoes in this formulation of the Vygotskian notion of the zone of proximal development; chapter 9 links Vygotsky and Sternberg to provide a theory-based approach to test administration.

### *3. The Componential Subtheory*

The basic unit of analysis in Sternberg's theory of intelligence is the information-processing component, namely, "an elementary information process that operates on internal representations of objects or symbols" (Sternberg, 1984, p. 281); this process replaces the unit of analysis in earlier theories, such as the "factor" or the "stimulus–response bond." An information-processing component may translate a sensory input into a conceptual representation, may transform such representations, or may translate them into motor outputs. The three properties associated with components are *duration* (time to execution), *difficulty* (probability of being executed erroneously), and *probability of execution* (the likelihood that an individual will arrive at an answer, whether or not it is correct).

Components perform three kinds of function: *Metacomponents* plan, monitor, and make decisions; *performance components* are processes used in task execution; and *knowledge-acquisition components* are used to learn new things.

*3.1 Metacomponents* are of special interest to neuropsychologists because of their explanatory power in getting into neuropsychology's "black box," referred to in previous typologies as the brain's third functional unit (Luria, 1973), the executive processes (Lezak, 1995), or the three components of adaptive behavior (Walsh, 1985). Sternberg defined metacomponents as "specific realizations of control processes that are sometimes collectively (and loosely) referred to as the 'executive' or the 'homunculus' " (Sternberg, 1984, p. 282): This is an endearing image, suggesting a little person, a homunculus, sitting in a cabin somewhere in the brain, steering the cerebral motor car along familiar or strange roads, accelerating down the straight and braking at the caution signs.

During automatized processing, metacomponents operate at the same level as process components and lose their executive character. This distinction at once draws attention to a puzzling phenomenon in high functioning individuals who sustain a frontal injury, namely, their complaint that "everything takes twice as long." This is because previously automatized process components now require the full attention of the global processor, thus ruling out the possibility of time-saving parallel functioning of a number of local processors simultaneously. Sternberg (1984) identified seven metacomponents (p. 282), not as an exhaustive catalogue but as exemplars of those most often encountered in intellectual functioning. Five of these are reviewed here.

*3.1.1. What Is the Problem?* The task here is to decide what the problem is that needs to be solved. For example, a group of intern clinical neuropsychologists at the doctoral level consistently "rigidifies" brain-injured clients against therapeutic interventions. Does this rigidification arise from incorrect joining with the client system? Or, is it the result of overly decisive information giving? Or, did the intake selection process pick a bunch of poor trainees?

*3.1.2. Selection of Lower Order Components.* For example, some syllogistic tasks are best solved by using spatial representation, and others by linguistic encoding; some tasks require that more time be allocated to encoding and less to stimulus combination and response. Selecting an inappropriate set of components will give rise to errors or inefficient performance. For example, an appropriate set of strategies on the Wisconsin Card Sorting

Test is flexible error utilization to move away from one sorting principle to another. Some clients, however, give all their energy to the deduction of a single bizarre and overcomplex sorting principle, such as "If there are two on top and one below, the next match must be with one below and two on top, then the next one must be four on top and none below . . ."; in fact, no cards exist that can match this fanciful principle.

*3.1.3. Smart Is Not Fast.* Conscious attention is a limited resource that must be shared among task components, so decisions on the allocation of attentional resources must be taken. Thus, it may be worthwhile to allocate attention in a way that tolerates slightly fewer errors by reducing speed in order to enhance performance accuracy.

Smart is not necessarily fast, argued Sternberg. And, as the review of cognitive competence in Africa at the end of this chapter shows, prudence and caution are more highly valued in many African societies than quickness. The assumption that smart is fast permeates our society and underlies most tests used to identify the gifted, according to Sternberg. However, it is not speed in itself that is critical, but rather the selection of an appropriate speed (see "Speed and Power"). For example, students are sometimes misled into believing that speed reading is more productive than regular reading, whereas it is reading speed flexibility that is the hallmark of skilled readers; study skills courses that exclude a speed reading component have produced better results than those that include it (Nell, 1988, chap. 6). As far back as 1924, noted Sternberg, Thurston proposed that "a critical element of intelligent performance is the ability to withhold rapid, instinctive responses and to substitute for them more rational, well-thought-out responses" (Sternberg, 1984, pp. 282–283).

Sternberg's own studies support the view that a reflective rather than an impulsive cognitive style produces more intelligent problem solving, although timed tests often force subjects to be impulsive; that better problem solvers spend rather more time on higher order planning than poorer problem solvers; that problem encoding is executed more slowly by more efficient problem solvers; and, finally, that faster readers give more time to reading passages on which they would be tested in detail and less time to other passages.

The conclusion is that although speed is important, most of the meaningful tasks individuals face in life do not require problem solving or decision making in the small number of seconds typically allocated for the solution of timed IQ test problems (Sternberg, 1984, p. 283). Inappropriate allocation of attention in the form of impulsivity and confusion are the hallmarks of the test performance of persons with compromised frontal function. For example, given the bicycle problem adapted from Walsh

Speed and Power

> A *pure speed* test is one in which individual differences depend entirely on speed of performance. Such a test is constructed from items of uniformly low difficulty, all of which are well within the ability level of the persons for whom the test is designed. The time limit is made so short that no-one can finish all the items. A *pure power* test on the other hand has a time limit long enough to permit everyone to attempt all items. . . . The test includes some items too difficult for anyone to solve, so no one can get a perfect score. (Anastasi, 1988, cited in Carroll, 1993, p. 444)

*Speed* and *power* are thus characteristics of tests, not individuals, who are judged rather in terms of rate of work and accuracy of response, also termed *level* of performance. (Carroll, 1993, p. 445). However, these logically distinct aspects of test performance cannot be distinguished under time limit conditions, because power is thus confounded by speed (p. 507).

Of special interest in the present context is Carroll's observation that "cognitive tests of level abilities given with time limits are seriously biased against individuals with low rates of test performance" (p. 508). This would, by definition, include individuals who are not test-wise (as defined in Sternberg's experiential archon), and those to whom deliberation is a more salient value than speed. What is one then to make of Thorndike's aphorism that "other things being equal, the more quickly a person produces the correct response, the greater is his intelligence"? Carroll correctly dismissed this as a societal judgment no longer supported by the data. The correlations between speed and level are too small and too variable "to justify any hope that rates of performance on elementary cognitive tasks could be used as indicants of intelligence level" (p. 508).

Matters are further complicated by the observation that reaction times cannot be taken as a univariate measure of apprehension or pure speediness, because the data reviewed by Carroll show that the more complex the task on which reaction time is measured, the longer the reaction time and the greater the correlation between reaction times and level abilities.

(1985, p. 238),[9] they say, with barely a pause for thought, "$25," simply subtracting the last two figures from one another. If asked, "Are you sure?" (a threatening question for a confused person), they think again for only a moment and, after impulsively adding the figures together, say "Oh no,

[9]A boy decided to sell his bicycle to a friend for $60. A few days later, the friend said he didn't like the bicycle and offered to sell it back to him for $40. The boy bought it back from the friend for $40, cleaned it up and polished it, and sold it again, this time for $65. How much did he make from beginning to end?

it's $105." The examiner might now say, "Explain that to me," and during the explanation a number of other answers, probably including the correct solution, will be offered, each no sooner given than retracted. This kind of impulsive helter-skelter method spills over into other aspects of problem solving. Thus, if all but one block on the more complex Block Design items has been correctly placed, such an individual is as likely as not to sweep away the entire assembly and start again, rather than to persevere with the fit of the last block.

Impulsivity is confusion's mask for uncertainty: Such performances support Sternberg's contention that a principle task of the executive homunculus is to allocate as much time as needed for correct problem solving.

*3.1.4.* ***Solution Monitoring.*** "As individuals proceed through a problem, they must keep track of what they have already done, what they are currently doing, and what they still need to do" (1984, p. 282). Good examples of different levels of working memory complexity can be drawn from arithmetic problems of increasing complexity. Tracking demands are low for single or two-operation computations, even quite tricky ones, such as working back from two thirds to the whole number, or saying how many items one can buy for a dollar if 12 cost 25 cents. They increase for items that are computationally simple, but involve a number of steps so that task solution requires the correct sequencing and storage of these steps, holding in mind the solution to one part of the problem while working on subsequent parts. For example, the computations required for the bicycle problem are derisory (subtracting 40 from 60 and adding the remainder to 65, which any primary school child can do); but, as with the Eiffel Tower problem,[10] the subject's task is tracking and storage rather than computation. In the visuospatial modality, the problem posed by the Complex Figure of Rey is precisely that defined by Sternberg—keeping track of what has been done, what is being done, and what still needs to be done. In the absence of such tracking, the design elements are ill-fitting and distorted. At a more concrete level, children on a formboard task may get stuck if they insert a piece at the wrong alignment, and in Sternberg's words, they cannot decide "whether to reperform certain processes that might have been performed erroneously, or to choose the best of the available options" (1984, p. 282). It is the failure to back away from an incorrect solution and try other options that is characteristic of cognitive rigidity and failure to make appropriate use of error recognition.

*3.1.5.* ***Sensitivity to external feedback*** is an executive prerequisite for effective task performance: Feedback must be understood, its implications recognized, and then acted on.

---

[10]"The Eiffel Tower is 300 m high. What is the length of one quarter of its height?" (Walsh, 1987, p. 145, citing Barbizet, 1970).

***Proliferation and Parsimony.*** The role of the central executive in regulating ongoing activity can be infinitely elaborated by naming additional metacomponents, as Luria did, for example, in *Higher Cortical Functions in Man* (1962/1980). The "riddle of the frontal lobes" arises in part from neuropsychology's failure to achieve a balance between proliferation and parsimony in the identification of executive processes. Parsimony produces an impoverished or rigidly schematic representation of the executive's functions; unlimited proliferation is chaotic; and a unifying cognitive schema that will systematize the diversity of frontal executive functioning without sacrificing richness has yet to be achieved.

*3.2* ***Performance components*** are processes such as encoding, representation, comparison, transformation, and response production: They are the bricks and mortar from which cognitive products are built, and are organized by stages of task solution, which in sequence are usually (a) encoding, (b) combination or comparison of stimuli, and (c) response. Each of these stages can be infinitely broken down into more elementary parts. For example, Sternberg showed that on an analogies task, the stage of stimulus combination-comparison can be separated into inference, mapping, application, comparison, and justification components. However, there is consensus (Sternberg, 1984; Verster, 1986) that an appropriate level of analysis is the information-processing component.

The decomposition of tasks into their component processes is an essential precondition for the diagnosis of cognitive difficulties, which is in turn a precondition for effective remediation. Sternberg (1986) gave the following example:

> Consider . . . the possibility of a very bright person who does poorly on tests of abstract reasoning ability. It may be that the person is a very good reasoner, but has a perceptual difficulty that leads to poor encoding of the terms of the problem. Because encoding is necessary for reasoning . . . , the overall score is reduced, not by faulty reasoning but by faulty encoding of the terms of the problem. Decomposition of scores into performance components enables one to separate, say, reasoning difficulties from perceptual difficulties. For purposes of remediation, such separation is essential. (p. 284)

*3.3* ***Knowledge Acquisition Components.*** Sternberg gave priority to the acquisition of knowledge, rather than to individual differences in the store of knowledge. Such knowledge may be both declarative (i.e., knowledge that answers the question "what?") or procedural (i.e., "how?"). Intelligence tests have a strong orientation to the knowledge already possessed by an individual: Vocabulary correlates very highly with the first principle intelligence factor, *g*, as do Wechsler subtests such as Information and Comprehension. This is not because knowledge is important for intelligence, argued Sternberg, but because a large knowledge base, which is what these tests measure, is an indirect indicator of the ability to acquire new knowledge.

In order to account separately for previously acquired knowledge and for knowledge acquisition skills, Sternberg presented subjects with very easy passages within which nonsense words were embedded. The task was to define the nonsense words by using the surrounding context. A precisely analogous task in the neuropsychological repertoire is Reitan's Word-Finding Test (Reitan, 1972). The subject's task is to discover the meaning of a nonsense word embedded in each of five sentences. To do so, clues in each previous sentence must be combined. Walsh (1985, p. 156) gave the responses of a patient with a left frontal tumor whose problem is an inability to transfer information from one sentence to the next:

1. Every house has a grobnick—no response.
2. Grobnicks are constructed in different shapes, but the purpose is the same—no response.
3. It would be quite uncomfortable to live in a house which didn't have a grobnick—"door."
4. A leak in a grobnick is usually detected during a heavy rain—"pipe or gutter."
5. Some people are very happy if they have food in their stomach and a grobnick over their head—"a hat or scarf." (p. 156)

## NORMALIZING THE EXOTIC

Luria's Uzbek expedition provides a dramatic example of the errors that result if the componential structure of a complex task is not understood. One of the first experiments his research team carried out in summer 1931 related to classical visual illusions. This was because he and Vygotsky wanted to demonstrate that Gestalt principles were the consequence not of brain structure but of culturally transmitted modes of perception. The results of these experiments so excited Luria that he sent Vygotsky the famous telegram cited in chapter 3, "The Uzbekis have no illusions!"

To researchers familiar with the African scene, this telegram will have a hollow ring. In Africa and Asia, neuropsychologists have been only too ready to impose their theory-based preconceptions on their subjects, and studies using European instruments and administration methods usually confirmed these opinions!

The contribution that careful componential analysis of abilities can make to an understanding of test results is well illustrated by the famous series of studies on pictorial perception in Africans. In 1960, Hudson produced data on adult illiterate Black factory workers in South Africa, who when shown perspective drawings of a hunter, an antelope, and an elephant in the background, "saw the pictures flat," and were unable to used depth cues to reconstruct the three-dimensionality of the scenes. Jahoda (1981) reviewed two decades of work in this area and drew two conclusions. First, the recognition of objects presented in pictures "is a human universal little affected by cultural variations" (p. 27). But, with regard to the perception of three-dimensionality, there is a strong correlation between the particular methods used by experimenters and the performance they elicited, indicating that "understanding pictures" had to be broken down into pictorial object recognition, pictorial depth perception, and a number of other component skills that still await research operationalization in order to determine the proportion of variance produced by each. Jahoda noted that only when this work has been completed will a pronouncement about the existence of a cognitive universal with regard to pictorial depth perception be possible. However, using the available methods, various investigators have shown that American children can handle depth cues from age 3, that Shona children in Zimbabwe are also able to perceive pictorial depth, and that secondary school pupils in Scotland, India, Ghana, Kenya, and Zambia correctly perceive depth. Jahoda concluded that "cultural differences in handling 3D representations in pictures have become negligible" (p. 40).

This careful progress from blanket statements about African mentality, presented in terms of a deficit model, to operationalization of the underlying constructs is also illustrated by Grant (1970, 1972). Grant, by the design of carefully tailored instruments, demonstrated that both spatial thinking and conceptual reasoning, until then thought to be dimensions largely absent from African thinking, could in fact be demonstrated in a group of largely illiterate mineworkers.

Grant's work marks the beginning of what has since become an inexorable trend in cross-cultural psychology to normalize the exotic. Today, concepts such as "African thinking" or "African mind" can no longer enjoy the status of cognitive constructs. But at the beginning of the century, the inhabitants of the New World were said to differ from the inhabitants of Europe and North America on a wide range of cognitive abilities.

Because a statement of the microgenesis of complex processes such as susceptibility to visual illusions or depth perception is now closer, the role played by culture in fostering or neglecting individual components of such complex skills at the level of executive processes becomes clearer, and assessment can move away from "the traditional opaque total score in which the separate contributions of accuracy, speed, persistence, and other characteristics are inextricably confounded" (Verster & Prinsloo, 1988, p. 556).

* * *

The utility of the triarchic theory in addressing these problems is by breaking away from the exclusively inner focus that has characterized mainstream psychology, and providing a framework within which neuropsychologists can on theoretical grounds accommodate culture- and education-related performance differences through a *contextual* sensitivity that acknowledges a culture's role in developing a powerful tacit knowledge for some abilities, but neglecting others that do not conform to that culture's definition of intelligence. This interpretation allows the neuropsychologist to ask whether a given neuropsychological test is likely to be an appropriate measure of cognitive competence in a particular individual's life experience. Sternberg's framework also allows assessment to take account of *experiential* factors by determining an individual's position on the novelty–automatization continuum, and to ask, "At this moment, what is the global processor doing? What aspect of task comprehension or task solution is engaging the subject?" Testing explicitly allows all subjects sufficient practice to acquire at least some automatization, and thus address themselves to solution rather than comprehension strategies.

The *componential* subtheory turns its attention away from person–environment interactions and returns to cognitive psychology's traditional inner focus, the individual's use of task solution strategies in problem-solving behavior. For all neuropsychologists, Sternberg's specification of the executive metacomponents, performance strategies, and knowledge acquisition components on which problem solving depends allows for richer analysis of intact and disrupted problem-solving behavior.

## III. The Piagetian Task Paradigm

Piaget's genetic epistemology is founded on a theory of cognitive universals. If such universals do indeed exist across cultures, then this would be convincingly demonstrated by the equivalent performance on piagetian tasks by children in the United States and Western Europe on the one hand and in culturally different settings on the other. These tasks have a narrower focus than psychometric tests, but are more complex than laboratory tasks and have a higher ceiling. From the wealth of available piagetian material, I have chosen to report one series carried out by Pierre Dasen of the University of Geneva (where Piaget was professor of child psychology from 1929 to his death in 1980) that is of particular interest to neuropsychologists, and has had an interesting outcome in southern Africa.

***Competence and Performance.*** Cognitive structures can be described in terms of *competence* and *performance* across three levels of cognitive sophistication: the preoperational, concrete operational, and formal operational (see Appendix 1). Dasen (1977) defined competence as an abstract and purely logical representation of what the organism knows or could implement in an ideal environment; thus, language is a universal human competence even in those born deaf and reared in isolation (Sacks, 1989). But performance requires a real device that can implement in real-life situations the knowledge or skill embodied in the competence.

In some cases, very little help is needed in actualizing a latent structure, that is, in moving a child from competence to performance. Dasen cited work with Eskimo children who moved from nonconservation to full conservation of quantity "after a mere exposure to other operational tasks," or very rapidly during the training phase; younger children required extensive additional training to make this shift (Dasen, 1977, p. 334). In older children, such "triggering" responses indicate that performance had been very close to competence (the bud, so to speak, had been about to burst into flower). But in younger children, the need for extensive additional training indicates that the competence existed, but performance had to be brought into being by appropriate stimuli. Under natural conditions, argued Dasen, "the competence for concrete operational structures is likely to be universal, but the way in which this competence is translated into spontaneous behavior is culturally determined" (p. 335).

In 1979, Dasen, Lavallee, and Retschitzki put this theory to the test among Baoul children between age 7 and 14 in a village some 200 km from Abidjan, capital of the Ivory Coast. Three piagetian conservation tasks (two of quantity and one of number), a compensation task, and a class inclusion task were administered to 14 children, with 8 nonconservers and 6 at the intermediate stage of conservation. The sequence of experiments was first a pretest, then four lengthy training sessions over a period of about a week, followed at once by a posttest on two of the tasks, and a further posttest on all five tasks a month later.

In general, the results demonstrated impressive generalization from one kind of conservation to another, and also to class inclusion (the child is shown 10 oranges and 3 avocados, asked how many of each there are, and then asked, "Are there more oranges or more fruit?"). Little evidence was obtained to support an "actualization of underlying competence" hypothesis, leaving the authors to conclude that competence was present, performance absent, but that the structures necessary for performance were created through training, which can be used to reduce the "developmental gap" repeatedly demonstrated between children from a Western, technological background, and those from traditional backgrounds in the developing countries. The gap does not represent the absence of competence.

But what evidence is there that some cultures differentially facilitate the emergence of skills neglected in other cultures? Is there an empirical basis for replacing the "deficit" interpretation of cultural differences by a "difference" interpretation, namely, that "differences in performance [are] accounted for by situations and contexts in which the competence is expressed" (Cole & Bruner, 1971, cited in Dasen, 1977, p. 335).

This question has been definitively answered by Berry (1966) in a comparison of perceptual skills among the Temne, a farming people of Sierra Leone, and the Canadian Eskimo of the Arctic coast. In essence, Berry hypothesized that because of ecological requirements and cultural bias, the Eskimo would be more aware of small detail than the Temne, and therefore more cautious about forming closure. And, for the same reason, the Eskimo would score significantly higher on measures of spatial ability. The results supported both hypotheses, indicating, as Berry concluded, "that ecological demands and cultural practices are significantly related to the development of perceptual skills" (p. 228). This brief summary does not do justice to the elegant quasi-experimental design of this study which, with related work, is described in Berry's (1976) *Human Ecology and Cognitive Style.*

Extrapolating from these findings, Dasen (1984) hypothesized that the Baoul farming people would do better than the Eskimo on a conservation task because people who produce, store, and exchange food will tend to value quantitative concepts more than hunter-gatherers, who might be expected to have little interest in precise quantitative comparisons. On the other hand, he anticipated that the Eskimo would do better than the Baoul on the conservation of horizontality (children must draw a line to show the water surface in pictures of bottles tilted at various angles) because "spatial skills and concepts [are] more valued in nomadic than in sedentary populations" (p. 11).

On conservation of liquids, all the Baoul but only 60% of the Eskimo children of the same age or older conserved fully; these findings are almost exactly reversed for the conservation of horizontality, showing that differences in the rate of development depend on which conceptual areas are valued in each culture rather than reflecting arrested development or lack of competence. Extending this series of studies, Dasen (1981) reported a time lag in the emergence of concrete operational quantity conservation for liquids, which is longest for the Eskimo of Cape Dorset, shortest for Western children in Canberra, Australia, and intermediate between these two for a sample of Baoul children. In addition, only some 60% of the Eskimo sample are able to conserve quantity between age 9 and 14. Similar asymptotes emerge on topographical tasks relating to spatial development in three different samples, one of Eskimo children from Cape Dorset, the second of Australian Aboriginal children from Hermannsburg, and the third of Ebri Africans from the Ivory Coast. This West African sample reaches asymptote at age 12. Irvine and Berry (1988b) commented that this plateau appears to cut off half the people in this group "from a whole array of mental tools and strategies that in western groups were in constant use by the age of 8 or 9" (p. 31). Dasen (1984) responded to this criticism: Asymptote does not indicate the presence of a cognitive deficit, but rather the lack of environmental press for development of the required performance (p. 411). In the light of the evidence from Dasen's other studies that training brings the necessary skills from latency to actuality, it would thus be incorrect to say that the nonperformers were "cut off" from the requisite skills.

***Training for Conservation in the Ciskei.*** A critically important predictor of test performance is years of formal education (Kendall et al., 1988). However, simple statements about level of education or schooling can be seriously misleading. Rogan and MacDonald (1983) deconstructed the notion of "schooling" in Africa, citing a characterization of Ghanaian schools that is widely applicable in the developing countries:

> Almost without exception, classroom instruction takes place at the formal level; children may see diagrams and photographs of objects they learn about, but the objects themselves are rarely available to the children. Instead, the child's role is primarily characterized by rote memorization of facts and events. (Arnold, Armah, & Cox, 1981, cited in Rogan & MacDonald, 1983, p. 313)

Accordingly, stratification of subjects by the amount of schooling they have had obscures wide differences in the quality of this instruction, and the emphasis it may or may not have placed on independent problem solving and critical thinking. These authors suggested that the conflicting findings on the relation between "schooling" and conservation skills should be understood in this light.

They went on to hypothesize that full conservation will develop if the instruction process allows for physical interaction with the apparatus, if there are opportunities to formulate experimental procedures and to reason through the results, and if there is exposure to cause–effect relations. These experiences were provided by the Science Education Project conducted in one of South Africa's former "homelands," the Ciskei, by the Centre for Continuing Education of the University of the Witwatersrand in the early 1980s. A total of 359 pupils participated in the project, of whom 62% attained concrete operational or late concrete operational problem solving; another 16% attained full late concrete operational thought. The respective percentages among 243 pupils not exposed to this intervention were 50% and 11% ($p < .05$). The conclusion is that if conservation skills are to be enhanced by schooling, then schools must consciously work at providing opportunities for physical and mental manipulation of the teaching environment and by helping to develop a cause–effect style of reasoning.

There is persuasive evidence in this series of piagetian task studies that the fundamental operations of intellect are invariant across cultures, but that—to use Vygotsky's metaphor—culture powerfully determines which cognitive buds will come into flower, and which will lie dormant until they are triggered by situational press.

## INTELLIGENT ACTS ACROSS CULTURES

The ecological sensitivity of the triarchic theory is primarily located in the contextual and experiential subtheories that build bridges between individual cognition and the cultural press of the environment by which the individual's thinking has been shaped. As Cole and his associates put it,

> Intelligence will be different across cultures (and across contexts within cultures) insofar as there are differences in the kinds of problems that different cultural milieus pose their initiates. In this sense, we must adopt the position of cultural relativists, such as Berry (1972) and Boas (1911) that no universal notion of a single, general ability, called intelligence, can be abstracted from the behaviour of people whose experiences in the world have systematically been different from birth in response to different life predicaments handed down to them in their ecocultural niche. (Laboratory of Comparative Human Cognition, 1982, p. 710)

### Cognitive Competence Across Cultures

In 1973, David McClelland advocated the assessment of competence rather than intelligence in a paper "that profoundly affected both the field of psychology and popular opinion" (in Barrett & Depinet, 1991, pp. 101–102). The term *cognitive competence* has since entered the vocabulary of cross-cultural psychology (Berry, 1984; Verster, 1986), linking in useful ways with the previous review of intelligence in cultural context.

Berry (1984) specified the steps that will have to be taken in order to achieve "a universal psychology of cognitive competence, one which is pan-human in scope" (p. 335), using cognitive competence as a generic and less prejudicial term than intelligence. He approached this task by adopting Irvine's call to psychologists to examine the ways in which laypeople use the word "intelligent" as a key to understanding "the ways in which society designates acts as intelligent" (Berry, 1984, p. 345, citing Irvine). Irvine (1969) analyzed 100 Shona proverbs that "represented ground rules for intelligent and purposive acts" (p. 98); the Shona word for intelligence, *ngware*, means cautious and prudent, especially in social relationships, leading Irvine to conclude that "intelligent acts are then of a conforming kind having primary reference to the affective climate of one's own relationships with the spiritual forces of the living and ancestral spirits of the kin group"

(p. 98). In Uganda, Wober (1969) replicated Irvine's study, reporting that the Baganda word *obugezi*, connoting both intelligence and wisdom, is thought of as slow and careful; the socialization of both boys and girls focused on maintenance of the social system rather than questioning or challenging it (Wober, 1974, cited in Berry, 1984).

Do Western concepts of intelligence have a good fit with African definitions? In Zambia, Serpell (cited in Berry, 1984) asked adults to rate village children by ranking them on the vernacular word for intelligence. These 42 children were then given a series of specially developed tests; the results correlated poorly with the adult ratings, suggesting that the tests failed to assess those child characteristics that their community saw as constituting intelligence (p. 348).

In a similar study, Dasen (1984) examined folk conceptions of intelligence among the Baoul of West Africa and the relation of these lay definitions to children's performance on piagetian tasks. The most salient aspect of intelligence in the view of Baoul adults was readiness to carry out tasks in the service of the family and the community, especially the willingness to carry out needed tasks without being asked. Following Mundy-Castle's (1983) distinction between social and technological intelligence, the Baoul also attached importance to skills of observation, attention, fast learning, and memory skills. But, when the children who had performed the piagetian tasks were ranked by adults on these indigenous conceptions of intelligence, most correlations were negative, leading Dasen to conclude "that the Baoul concept of intelligence is basically different from concrete operational reasoning [and] that spatial skills are not valued in this culture" (pp. 428–429).

In essence, these studies confirm Irvine's view that intelligent acts are of a conforming kind. However, with increasing Westernization, there is a marked shift away from slowness and carefulness as the criteria of intelligent behavior toward the desirability of quickness. Mundy-Castle (1983) captured this movement in the distinction he drew between "social intelligence" and "technical intelligence." Berry reported further studies of lay concepts of intelligence in Guatemala, Malaya, China, Taiwan, Australia, and the United States.

Can this large body of knowledge about intelligent acts in different societies be converted to an integrated universal theory of intelligence? Such frameworks have been proposed (Berry, 1984, Figure 2), but the goal is unlikely to be realized: It would be necessary to demonstrate the ecological validity and equivalence of a huge set of derived etics across the full range of human cultures. "In practice, this implies a research programme of unrealizable scope" (Verster, 1986, p. 28).

The notion of *cognitive* competence has a significant additional difficulty in that it devalues the range of emotionally based adaptive behavior: tact, considerateness, conversational pragmatics, and intact affectivity (Damasio, 1994; Prigatano, 1991). Neuropsychologists, in particular, require a wider definition of competence that includes all of adaptive behavior, and not only its intellectual aspects. It is interesting in this context that Sternberg, Conway, Ketron, and Bernstein (1981) found that a "social competence" factor emerged when laypeople rated intelligent behavior.

Can competence testing live up to the hopes it aroused in the 1970s? Barrett and Depinet (1991) concluded that it has not been more successful than traditional intelligence and aptitude testing in predicting job success or in avoiding prejudice to minorities; there is little evidence that "competency testing has the potential to make a unique contribution to the field of testing" (p. 1021).

However, in culturally different settings, the need for alternative approaches to the quantification of ability remains urgent. *Competence* as a construct is heir to the conceptual ills of intelligence: One reason for this failure is that the cognitive content given to theories of competence has focused exclusively on the culture-laden executive processes, which are only one component of intelligent behavior. On the other hand, the measurement of *cognitive potential* (chap. 9) may offer a viable alternative method of ability assessment.

## A CASE STUDY AS A FOOTNOTE: ENGLISH–AFRIKAANS IQ DIFFERENCES

As a final step in broadening the canvas, it is useful to consider the marked intelligence score differences between White English- and Afrikaans-speaking South Africans. There are no race differences between these groups; they share a cultural heritage, are brought up within a unitary state, and are compelled to acquire a working knowledge of one another's language; their socioeconomic status is equivalent, and so is their education. The force of the following material is thus to show that even very subtle cultural differences that in this case relate only to social values and norms can have a substantial effect on IQ scores, and such scores cannot therefore reliably reflect an innate, universal "intelligence." Let us see what might be learned from this state of affairs that is helpful in understanding IQ differences between White and Black Americans, and other intergroup disparities.

An introductory word about South Africa's two White tribes will be helpful. Afrikaners are descendants of the Dutch, who settled the Cape from 1652. English settlement began with the first British occupation of the Cape in 1806, and expanded greatly with the arrival of a large British settler group in 1820 in the wake of the Napoleonic wars. To escape British rule, groups of Afrikaners moved north and east into South Africa's interior, fighting a series of bloody skirmishes with African tribes. This Great Trek has been mythologized (Hofmeyr, 1991) as the birth of the Afrikaner nation, simultaneously facing a White enemy, the British, and African tribes. After the discovery of diamonds and gold in the late 19th century, the Afrikaners fought and lost a bitter war of independence against the British from 1899 to 1902.

Afrikaner ascendancy in South African politics began with the election victory of the Nationalist Party—the architect of Apartheid—in 1948, and was symbolically consolidated by South Africa's withdrawal from the British Commonwealth in 1956. The late 1950s were thus marked by the consolidation of Afrikaner political power, the extension of this power into heavy industry, growing government control of universities and other state-funded research institutions, and a repressive political climate.

Against this background, it is hard to imagine a more emotional and politically sensitive issue than a report in 1959 that Afrikaners were of lower intelligence than English speakers.

## A Difference of Six Scale Points

Standardization of the Wechsler–Bellevue Intelligence Scale for White South Africans began in the 1940s, and used a sample of 1,500 English-speaking and 1,500 Afrikaans-speaking South Africans ranging from age 18 to 59. In 1959, Biesheuwel and Liddicoat reported that a reanalysis of data from a sample of 305 English-speaking and 299 Afrikaans-speaking subjects from this norm group showed that English speakers scored on average 6 scale points higher than Afrikaans speakers (nearly half a standard deviation, significant at the 1% level). When the sample was stratified by socioeconomic status (SES), no significant difference was found at the lowest SES category, which, argued the authors, showed "lack of responsiveness due to low intelligence and an environment catering for little more than the necessities of life [which] would leave little scope for differential cultural development" (p. 5). At the highest SES level, the difference, although still significant, was lower than in the intermediate SES range, implying that developmental conditions might converge for both language groups at the highest and lowest income levels.

Stratifying the sample by both socioeconomic status and subtest performance showed that English and Afrikaans speakers had virtually equal scores on the Arithmetic subtest, because, argued the authors, "arithmetic is a subject in which all scholars are drilled alike, and in which a specific cultural effect is therefore least likely to occur" (p. 8). Block Design and Digit Symbol Substitution, although supposed to be free of cultural bias, also yielded a significant English-speaking advantage in the highest and intermediate SES categories, suggesting a difference in fluid intelligence levels.

Biesheuwel and Liddicoat (1959) concluded that these persistent differences in favor of English speakers pointed to "a more general and fundamental influence, operative in all the tests, and most probably affecting the level of *g*" (p. 11). Given the political climate in South Africa in the 1950s, their conclusions are remarkably hard-hitting:

> The possibility must at least be considered that the lower stimulus value of the Afrikaans-speaking environment (in terms of parental education and interests, diversity of aspirations, material culture in the home, frequency and intensity of contacts with people, ideas, and intellectually challenging situations) has drawn out the intellectual potentialities of its members to a lesser extent than the English-speaking environment. (p. 12)

Partly as a result of this publication, Biesheuwel's position as director of the South African National Institute of Personnel Research came under siege, and in 1962 he resigned his post. Biesheuwel died in 1991 at age 83, and his son-in-law, John Verster, wrote in his obituary notice that the Afrikaner establishment had in the 1950s read into Biesheuwel's work "an attempt to denigrate Afrikaans South Africans as intellectually inferior" (Verster, 1991, p. 269), and responded defensively. In 1960, for example, Langenhoven (1960) argued that the conclusions drawn about the comparative "intellectual potentialities" of different cultural groups were premature (p. 152), and suggested that the difference in test performance could be ascribed to a lack of test sophistication rather than performance differences in real-life situations.

### The Continued but Diminishing Gap

Reanalysis of the South African Wechsler–Bellevue data by Verster and Prinsloo (1988) shows a steady increase in the magnitude of the difference in favor of English speakers from age 45 onward. This is not because of differential aging effects on intelligence, but rather the result of differential cohort experiences: "The cultural distance between the two populations is likely to be greater in the older cohorts [born in the decade of the 1890s] than in the younger" (p. 541).

Another large investigation of English–Afrikaner test score differences was a Human Sciences Research Council survey comparing 21,000 English-speaking with 41,000 Afrikaans-speaking children who were in Std 6 (the eighth year of formal schooling) in 1965, and retesting these cohorts in 1967 and 1969. In all comparisons, English speakers outperformed Afrikaans speakers, and more markedly so on the fluid intelligence nonverbal scale of the New South African Group Test. Among Std 6 pupils, English speakers fared significantly better than Afrikaans speakers on tests of paragraph memory, memory for words and symbols, number ability, and reasoning.

This gross difference masks a far more interesting process, namely, that in comparison with the Biesheuwel and Liddicoat data there has been "a progressive decrease in ability score discrepancies with increasing cultural convergence over successive generations of white South Africans" (p. 544).[11]

In an extension of this series of studies, Verster (1974) found significant differences between English- and Afrikaans-speaking graduate research scientists at the Council for Scientific and Industrial Research, again in favor of English speakers. Factor analysis of the six tests he used showed that five loaded on what he called a common reasoning factor, akin to $g_f$, in which inductive and deductive variance merged, whereas the sixth loaded on spatial reasoning for English speakers, but on both space and reasoning for Afrikaans speakers (p. 546). To what can these continued differences in the 1970s be attributed? Verster suggested that they are due "to stylistic differences in approach to intellectual tasks [and differences] on personality dimensions such as over-regimentation and conservatism, authoritarianism, and ethnocentrism" (pp. 546–547). In a subsidiary analysis of the same data, Sussenguth (cited in Herrnstein & Murray, 1994) found that a battery of self-report items showed higher scores for the Afrikaans speakers on such dimensions as "rigidity versus versatility in thinking, ideational conformity versus ideational independence, and low performance potential versus high performance potential" (p. 546), thus confirming that there are indeed underlying stylistic differences in approaches to problem-solving tasks:

> These findings suggest the need for extreme caution when wishing to infer equivalence of psychological meaning in tests [even] when psychometric criteria for equivalence have been met. Mean population differences on ability tests are at least as likely to reflect the stylistic or other differences in approach to the tests as differences on the presumed underlying "ability." (Verster & Prinsloo, 1988, p. 554)

---

[11]A similar process has been at work in the United States. Data from the National Assessment of Educational Progress (NAEP) show that over a 20-year period, there has been an overall narrowing of the Black–White IQ gap equal to .28 of a standard deviation, or about 5 IQ points. Some age-specific subject gains are much higher. For example, for 17-year-olds, reading scores improved by .44 of a standard deviation over this period. Speculating on the reasons, Herrnstein and Murray (1994) noted that the quality of Black schooling has improved, nutrition and health care have improved, travel opportunities have increased, and media exposure has reduced the impact of environmental differences.

## *THE BELL CURVE* IN CROSS-CULTURAL PERSPECTIVE

It is untrue to say that the explanation for these differences is either genetic or cultural. Ordinary people do not have to go to college to know that bright parents usually have bright kids, and that siblings have similar intellectual endowments; there is a proverb in English and many other languages, "The apple never falls far from the tree." There is an overwhelming commonsense and scientific case for a large genetic contribution to intelligence, just as genetics contributes to every other characteristic that marks individuals as members of the species, as the children of their parents, and as individually unique.[12]

There is also an overwhelming case for the cultural moulding of intelligence. Herrnstein and Murray, the arch-nativists, acknowledged the effects of school quality, nutrition, health care, travel opportunities, and media exposure on IQ; their comment on "the gaping cultural gap" between rich and poor was cited earlier. Jensen (1969) pointed out that if the first IQ tests had been devised in a hunting culture, " 'general intelligence' might well have turned out to involve visual acuity and running speed, rather than vocabulary and symbol manipulation" (p. 14). But even more telling, because it comes from a different place and time, is the attribution Biesheuwel and Liddicoat (1959) made nearly 40 years ago: that Afrikaners score lower on IQ tests as a result of "parental education and interests, diversity of aspirations, material culture in the home, frequency and intensity of contacts with people, ideas, and intellectually challenging situations" (p. 12).

The data here and in chapter 2 show that there is a general tendency for developing country populations to score lower on psychological tests than Western subjects; some enclave cultures embedded within larger populations—African Americans, Hispanics, Jews, English-speaking South Africans—also score lower or higher than the population mean. The lesson to be learned from close on a century of cross-cultural psychological assessment is that until language proficiency, educational quality, test-wiseness, cognitive style, and socially mediated definitions of what it means to be smart have been shown beyond any reasonable doubt to be equivalent for the groups whose scores are being compared, then score differences cannot be attributed to genetic differences. Scarr (1978) cited evidence "that black children are being reared in circumstances that give them only marginal acquaintance with the skills and knowledge being sampled by the tests" (p. 335; see also Mercer, 1984). For these and other lower scoring

---

[12]It is "ridiculous to suppose that abolishing intellectual measurement will revolutionize anyone's life chances," or that "biological diversity must be denied to defend universal civil liberties" (Scarr, 1978, p. 327).

groups, the test as given may not be the test as received, invalidating score comparisons.[13]

Similar reasoning applies with regard to groups that do better than the population average: Genetic attributions for the higher scores of Jews and Asians are supect until due weight has been given to the foundational assumptions of the culture that determine achievement motivation, among these the value placed on speed and classroom learning. So, even after careful validation and norming, psychological tests are a chancy way of getting through to the underlying construct of adaptive capacity in subjects from other cultures.

But, despite these daunting obstacles, both construct validity and valid norms for populations outside Western Europe and North America are within reach, as the next chapter argues: This will take time and money, but there are no theoretical obstacles.

[13]The review of validity issues in chapter 5 further shows that predictive validity—the indicator on which Herrnstein and Murray (1994) based their conclusion that test bias does not account for Black–White test score differences—is compromised by validity generalization (Cronbach, 1983, in Verster, 1986), and is therefore likely to be a misleading indicator of construct validity.

*Chapter* 5

# Constructs, Norms, and the Problem of Comparability

For decades, construct validity has been the bugaboo of psychologists working with people from other cultures. Neuropsychologists in particular have learned to their cost that a report that is the last word in qualitative interpretation will be dismissed by the crusty voice of methodology: "Yes, doctor, indeed, but what about constructs?" The more psychometrically sophisticated the setting, the more difficult it is to move beyond the construct validity blockade.

Researchers and clinicians respond to these problems rather differently. Researchers take the purist view, say solemnly, "construct validation is the *only* thing" (Carroll, 1983, p. 213) and insist that until the underlying meaning of the test has been determined, norms for that test will be meaningless. Clinicians, on the other hand, badgered by their clients to say something prognostically useful, cannot retreat into an ivory tower. On the one hand, they face imperious real-world demands from clients, families, rehabilitation professionals, and, increasingly, from criminal and civil courts, to provide accurate information about the previous and current abilities of clients who may have sustained brain damage. On the other hand, the material in the previous chapter shows that the use of Western tests with clients from other cultures can be so misleading that a cultural veto should be imposed on the use of such test scores.

A vicious circle now develops. In the first place, cross-cultural psychology has made little progress on the construct validation of psychological tests in the developing countries. Without constructs, there are no norms, and because test constructs are in doubt, psychological research institutions in the developing countries have felt unable to mount large-scale norming projects. Without norms, the forensic status of psychological and neuropsychological tests is in doubt, and the courts are increasingly driven to fall back on medical determinations as to whether or not a plaintiff sustained a brain injury. But neither a neurologist's determination of neurophysical normality, or a Glasgow Coma Scale in the mild brain injury range (between 13 and 15; Nell, 1997b), can rule out the existence of subtle but devastating cognitive and behavioral deficits (Benton, 1989; Blakely & Harrington, 1993; Leininger, Gramling, Farell, Kreitzer, & Peck, 1990; L. F. Marshall & Ruff, 1989).

The history of cross-cultural psychological assessment adds another dimension to these problems. Although the idea of assessing truly emic abilities has been actively canvassed since the 1960s (Irvine, 1966, 1969),[1] not a single method of assessing an ability indigenous to a specific culture has moved from the cognitive laboratory to the clinician's consulting rooms; similarly, mainstream neuropsychological assessment seems content with its Western focus and indifferent to the needs of psychologists and their clients outside the countries of Western Europe and North America.

The resulting distress of clinicians in a developing country situation is well-captured by Tollman and Msengana (1990), who attempted to adapt *Luria's Neuropsychological Investigation* (LNI; Christensen, 1974) for use with Zulu speakers in Natal, South Africa. They wrote that the validity problems that arose led some of their critics "to intellectualise the feasibility of the project out of existence; thus the *status quo* would have to remain until some future time. . . . We do not share this view; we have to do something to resolve the dilemma" (p. 20). Unfortunately, the expectation that cul-

---

[1]Cross-cultural psychologists make a useful distinction between the *etic* (or universal) and the *emic* (or particular). These terms derive from the words phonETICS, which is the study of the general and universal aspects of sound in language, and phonEMICS, the science of sounds that are meaningful and employed within a single linguistic system (Berry, 1984; Pike, 1954). An *imposed etic* (Berry, 1989) is a Euro-American given imposed blindly on a set of phenomena that occur in other cultural systems. For example, in studies of general intelligence across cultures, the assumptions are most often universalist, namely, that intelligence exists among all people and the only question is how much of it there is in a particular people. Such examinations are usually carried out by "adapting" existing intelligence tests, which means translating them, doing only those modifications or additions necessary to obtain data, and then administering them (see "The Perils of Adaptation").

A *derived etic*, on the other hand, emerges from a cross-cultural range of phenomena and is empirically derived from the common features of these phenomena. Irvine's series of studies of indigenous concepts of intelligence (reviewed in Berry, 1984, and Irvine & Berry, 1988a, pp. 47–49) illustrates the classic method of establishing derived etics. But, as other cross-culturalists have wryly noted, the etic is wily, and one cannot assume that a particular study has succeeded in snaring a new member of this species: "There may always be a deviant emic lurking in the shadows, ready to upset an otherwise productive hunt for regularities in human behavior" (Trimble, Lonner, & Boucher, 1983, pp. 259–260). A paradox of cross-cultural psychology is that in order to develop a universal theory of cognitive competence, one must first adopt a stance of *radical cultural relativism*—that no psychological universals can be assumed across cultural systems (Berry, 1972, p. 78).

turally mediated differences will be eliminated by "adaptation"—providing an accurate translation, and substituting local content for that in the original—rests on the unspoken assumption that there are pan-human cognitive universals, and therefore that all cultural effects reside in language.

But, dismissing the problem of construct equivalence as if adaptation is all that is needed gives rise to unexpected outcomes. Here, the authors noted with puzzlement that when they asked their subjects to place a circle in a given position within a parallelogram (a seemingly simple task that in fact draws on a host of education and culture-dependent skills), their subjects drew the circles haphazardly:

> They sometimes spent substantial amounts of time staring at the figures, not knowing where to begin. They seemed confused by the apparently unfamiliar designs. The Rupp's test (completing a pattern that resembles a honeycomb) also appeared to present difficulties. The problem here could have been in perceiving the regularity of the design. (p. 21)

The pressure to provide usable assessment instruments for clients in settings other than Europe and North America is real; but constructs are also real, and measures of nonexistent constructs are just not going to work.

## THREE KINDS OF COMPARABILITY

Progress toward construct validation begins when neuropsychologists are able to convince research funders that score gathering is not a bolt-hole to escape the duty of construct validation, but good science: "Scores are the input to the determination of cognitive structures" (Carroll, 1993). The purpose of this chapter is therefore to move beyond ritual lamentations by laying the foundations for the practical solutions suggested in Part III. To do so, two theoretical issues and one practical issue are reviewed. The first of the theoretical issues is to define the different types of comparability; the second is to distinguish between strong and weak construct equivalence, based on a distinction between culture-bound executive processes on the one hand, and hypothetically universal performance processes on the other (Verster, 1986). Although theoretically appealing, it is by no means clear that this distinction will be helpful to cross-cultural assessment. The practical issue is to describe the steps that must be taken to arrive at a compendium of construct-based norms for culturally different clients.

### Predictive Comparability

As the review of *The Bell Curve* in the previous chapter showed, predictive comparability is evaluated in terms of an external criterion, often achievement at school or university. However, predictive power cannot be used to infer construct equivalence. "Adapting" Western tests of brain damage requires sweeping validity assumptions that are seldom justified. Moreover, such adaptation might yield tests that appear to be good predictors not because of construct equivalence, but because the phenomenon itself is so robust that any test at all will be diagnostic. What is at work here is the worrisome phenomenon of validity generalization:

> Almost any cognitive test can be shown to have validity for almost any performance criterion, . . . especially in the case of illiterate or semi-literate samples in Africa, where good predictive abilities are found with virtually any psychometric test. (Cronbach, 1983, in Verster, 1986, p. 7)

Factors extrinsic to the test task (and therefore to the construct purportedly measured by the test) account for the positive correlations between test scores and the criterion. On an aptitude test to select participants in a training course, for example, this extrinsic factor might be good language comprehension skills that help the client understand the test instructions and later profit from the tuition given on the course.

Validity generalization is of course the reason that most neuropsychological tests seem to "work" in other cultures, even if construct validity is weak. In consequence,

> trying to erect a theory of the construct meaning of tests on the basis of predictive validity is like attempting to define the true colour of a chameleon; with every new criterion, the operational definition of a test must be revised. . . . The empirical relation between test and criterion implied in such validity coefficients cannot be interpreted in the absence of theory. (Verster, 1986, pp. 6–7)

The chameleon metaphor is strictly accurate. The same tests administered to the same clients will very likely predict scholastic aptitude, brain damage, or "trainability" with equal facility.

**Construct Comparability**

A test has no norms until its construct validity has been established. It can have scores, and even impressive-looking standard scores, *but it has no norms.* Once the underlying constructs are known, scores become norms. Failure to meet this condition leads to injustice, because scores that are misleadingly low deny clients jobs for which they qualify; if they are too high, clients are deprived of the compensation to which they are entitled. But, if construct validity has been established, then the psychologist can with reasonable certainty say to parents, teachers, employers, or a court of law that the client's performance on this or that test was in the normal, superior, or impaired range, and go on to explain what the tests in question measure and what they predict.

Construct comparability is thus the most fundamental type of comparability. Unfortunately, it is also the most difficult: Abstractness is the essence of a construct, which is defined as "a measure of some attribute or quality which is not operationally defined" (Cronbach & Meehl, 1955, p. 282). Depression is a construct, and so is analogical reasoning; but reaction time and tapping speed, which can be directly timed or counted, are no more constructs than the width of this page in millimeters. However, abstract though they may be, constructs are measurable by psychometric tests or information-processing probes that accurately access the construct in question.

From within the information-processing paradigm, the easiest way to conceptualize constructs is as metacomponents in their own right. They are, like all metacomponents, accessible to consciousness and have a representation in language (see "Weak Construct Equivalence: Executive Processes" later). Chiefly, however, each of these *metacomponent constructs* is a mental representation of a task or a behavior in the real world.

Problems relating to constructs multiply as one moves across cultures. In the first place, before construct validation can begin, the construct itself must be shown to exist as an identifiable entity in the heads of the target group—otherwise the test cannot elicit it. A scale to measure asceticism will work well in a monastery, but not among hunter-gatherers living in a harsh environment.

***An Example of Construct Invalidity.*** The Profile of Mood States (POMS) provides a striking example of the breakdown of construct validity assumptions. The POMS was included in the WHO–NCTB, a supposedly universal assessment instrument, because of its face validity. This is an all-too-frequent error by test compilers, who assume that if a procedure seems to draw on cognitive universals, or appears to make no literacy demands, it can safely be included in batteries for use in many different settings. On these spurious grounds, the POMS certainly appears to be a good candidate for cross-cultural use. However, Chinese investigators have questioned its validity (Liang, R. K. Sun, Y. Sun, Chen, & Li, 1993) and a recent multicenter study found that the POMS was not usable among semiliterate Nicaraguan subjects (Anger et al., 1993). Mokhuane (1997) applied the 65-item POMS to 58 semiliterate Tswana- and North Sotho-speaking South Africans and found that of the six factors, five intercorrelated at values between .70 and .91: *tension-anxiety*, *depression-dejection*, *anger-hostility*, *fatigue-inertia*, and *confusion-bewilderment*. She thus concluded that these factors (reported as factorially independent scales by McNair, Lorr, & Droppelman, 1971, 1992) lacked construct validity for her subjects, and treated all five as a global "general mood factor." The sixth, *vigor-activity*, clearly a facet of behavior rather than mood, and thus more concrete and readily understood, was independent of the other five.

For our group's studies in South Africa (Brown et al., 1991; Nell et al., 1993), the scale was translated and back-translated, where necessary, substituting phrases for single-word mood descriptors (e.g., translating "bushed" as "Are you feeling tired today?"). Subjects in the second and larger of the two studies were middle-aged (46 years, *SD* = 9.8) factory workers with a mean of 6 years schooling (*SD* = 3.1) whose home language was Zulu or South Sotho. Two difficulties arose. First, the translators complained that the number of synonyms in these African languages was inadequate for all of the synonyms used on a given dimension of the POMS. For example, the synonyms the scale uses for "fatigue" are "worn out," "listless," "sluggish," "weary," and "bushed"; for "angry," these are "peeved," "grouchy," "annoyed," "ready to fight," and "bad-tempered." Despite the best efforts of the translators, subjects would respond with irritation (anger, peevishness) to the synonyms, saying, "I have already given you the answer, why are you asking me again?"

Second, subjects had difficulty with the instruction to describe "how you have been feeling during the past week, including today." Test-wise subjects have no difficulty in understanding that the examiner is after a trait rather than a state; our subjects, on the contrary, would respond in terms of their current state. For example, asked if they were feeling weary, subjects would ask, "Do you mean when I got up this morning? I was feeling very tired. But on the way to work I felt better. Now [at 3 pm] I am tired again. Must I tell you for this morning or for now?"

Similar difficulties for much the same reasons arose in the same study with the Subjective Symptoms Questionnaire, which also forms part of the WHO–NCTB. Questions such as "Do you often forget things?" or "Do you often have a headache?" were interpreted in unexpectedly literal and concrete fashion. Nonetheless, this test was administered to all 68 subjects, but scores correlated positively with neuropsychological test scores. In other words, the more symptoms our subjects had, the better they did on neuropsychological testing! Clearly, the constructs the test designers had in mind were not accessed by this questionnaire.

## Score Comparability

Scores cannot be compared until construct comparability has been shown to exist. An error that derives from the universalist conviction—to which neuropsychologists are peculiarly prone—is to report norms for tests that measure unknown constructs. The Trail Making Test, illustrated in Fig. 4.1, is again a useful vehicle for this discussion. Recall from chapter 4 that the difference between Part A and Part B is that in the first, the client follows the number series, drawing a line from 1 to 2, from 2 to 3, and so forth, whereas in Part B, the number and alphabet series are simultaneously followed so that these two entrenched series conflict with one another at each step: "If 1 then 2, no, I must say A!, if A then B, no, it's 2!", and so on. The global construct Part B measures is stimulus resistance; in Western societies, it can be taken for granted that almost all subjects will have had 10 or 12 years schooling, and the stimulus resistance construct will be present because both these series will be fully entrenched.

Matters are otherwise for semiliterate subjects. Table 2.1 confirms that for such individuals, execution times for saying the alphabet and counting from 1 to 10 are prolonged, so even if the client can recite the letters of the alphabet in the right sequence, this will probably be slow and hesitant, as will counting, showing that neither of these two series is entrenched. Accordingly, the client will feel no conflict in moving from 1 to A and A to 2, so whatever the test is measuring has nothing to do with stimulus resistance. So, if the scaled scores of the client on Parts A and B are more or less equal—as is very likely—it would be absurd to conclude that "despite his low level of education and severe brain injury, Mr. XYZ showed no evidence of a working memory problem on the Trail Making Test." A more reasonable inference is that the neuropsychologist who came to this conclusion has impaired abstract reasoning ability!

## WEAK AND STRONG CONSTRUCT EQUIVALENCE

If, as suggested in the introduction to this chapter, a way could be found of distinguishing culture-bound aspects of human performance from its pan-human aspects, then the task of identifying the constructs that determine how tasks will be comprehended and executed would be greatly simplified.

Sternberg's triarchic theory offers an approach to this analytic task by distinguishing between the context of intelligence and its components. Mediating between these two facets of intelligence is experience: Context determines what is novel for an individual in a culture and what is familiar, and as a result what will be automatized and what will remain under the control of the global processor. In this way, context ultimately determines the content of the metacomponents and the ways in which they will drive performance.

Verster (1983, 1986) extended this idea by arguing that the metacomponents are the locus of culturally determined differences in cognition due to differences in socialization, learning, cultural habits, and ecological press. Put differently, a culture-laden executive dictates differing performance styles. Performance processes such as encoding, registration, comparison, and transformation, on the other hand, are characterized as mechanistic, basic elementary processes that are likely to be free of cultural influences and thus to constitute universals in human cognition—although there will nonetheless be individual and group differences in efficiency of execution.

On theoretical grounds, weak and strong construct equivalence can now be distinguished.

### Weak Construct Equivalence: Executive Processes

Although all three aspects of Sternberg's componential theory are affected by environment and experience, it is the metacognitive "drivers" of the processing system that are most heavily governed by environmental press and cultural values. In other words, it is the homunculus that decides whether to emphasize speed or deliberateness, what to automate and what to leave to the global processor. To the extent that executives are mental representations of tasks and behaviors determined by socialization, they are accessible to consciousness: Unlike performance processes, they can talk. Moreover, through introspection executives are able to give some account of themselves in conversational inquiry. But with unsophisticated subjects, this will usually have to be mediated by the examiner.

In pilot testing, the Performance Probe Battery described in chapter 9, we talked each of our subjects through the tests, asking them to tell us what they were doing, what was easy and what was difficult, and what they thought we were testing. In effect, we were educating the clients' executive to view the test through the test-maker's eyes, and to adapt their own performance to those expectations. In Appendix 3, the obverse of this fascinating process is described, namely, a method of tapping into the client's executive processes in order to construct test instructions that have the best fit with the client's cultural milieu and customary expectations.

### Strong Construct Equivalence: Performance Processes

Attention can now be turned away from the executive, whose nonequivalence across cultures is taken for granted, and turned toward performance processes, the elementary building blocks of cognitition, to determine how much each is affected by—or immune from—group and cultural differences in speed and efficiency. On theoretical grounds, this will allow strong inferences about construct equivalence to be drawn for two reasons. First, executives cannot themselves be directly measured. The effects of their style and priorities must be inferred by working backward from performance parameters. Performance processes, on the other hand, have motor expression and can be measured along parameters such as latency to response, response execution time, total correct and incorrect responses, and hence indexes of speed and power leading to inferences about response style.

Second, and more important, performance processes are the final common path through which the products of cognitive activity are expressed. Although one may speculate about the effects of context on the executive, and explore such speculations in conversational introspection with bearers of the culture, the quantifiable effects of culture on cognitive processes are directly measurable only as performance outputs. Performance processes, unlike executives, are mute, having no language beyond their motor products, but this measurability is their inestimable advantage, allowing direct quantitative comparison between the response parameters of bearers of different cultures and thus determination of whether the underlying metacognitive drivers (the mental constructs) are or are not equivalent in these cultures. Strong construct equivalence might thus be established.

On theoretical grounds, the most intriguing by-product of this analysis is that it identifies what seems to be a fundamental flaw in the methods hitherto pursued for the construction of a universal account of human competence: The elements of indigenous intelligence reviewed in chapter 4 are uniformly at the level of executives rather than performance processes, and as such are unquantifiable. Theory therefore suggests that the best way to bypass the effects of culture would be to access performance processes directly.

***Searching for the Building Blocks of Cognition.*** As noted earlier, this is an attractive prospect, and it remains an important research agenda for psychologists concerned with the assessment of culturally different people. But at this time it is unlikely to bear practical fruit: If a task is cognitive, it cannot be elementary; if it is elementary—in the sense that the reflex loop appears to be elementary (although of course it is not: even reflexes are modified by the executive)—it is of no interest to psychologists, and will at best make a marginal contribution to psychological assessment.

Performance processes, the building blocks of cognitition, are analogous to elementary cognitive tasks (ECTs), about which Carroll (1993) made the same point:

> Some authorities dislike the expression, "elementary cognitive task" because in their opinion it wrongly suggests that cognitive tasks can be elementary and simple when in fact they are probably very complex, or that tasks can be decomposed into simple components when they are actually not amenable to such decomposition. (p. 28)

Summarizing the factor analytic data, Carroll noted that the evidence correlating ECTs with psychometrically defined cognitive abilities is problematic (p. 506). It is of particular interest that the size of the correlation coefficients (typically small but significant) increases as the information-processing demands of the task become more complex (p. 508). Again, this suggests that theory notwithstanding, psychological and neuropsychological assessment would for the time being at least do better to use established psychometric rather than innovative information-processing tasks. This is therefore the approach that I have followed in assembling the core test battery proposed in chapter 10.

## FROM SCORES TO CONSTRUCTS TO NORMS

> Clinical neuropsychologists are always starving for good normative data. (Fasteneau & Adams, 1996)

Norms, both good and bad, are plentiful. The word occurs 2,108 times in the *PsychLit* database from 1990 to 1996, and is paired with neuropsychology 88 times. For sub-Saharan Africa, a search commissioned by the South African Clinical Neuropsychology Association located 475 references to test scores or norms (Nell & Kirkby, 1997).[2]

But good norms are indeed rare. Two recent reviews of published norm compilations show the disappointment and irritation of practitioners when basic methodological requirements are not met (see also Nell, 1997a). Artiola i Fortuny (1996) dismissed the norms in *Neuropsychological Evaluation of the Spanish-Speaker* (Ardila et al., 1994), as "unreliable if not worthless" (p. 234) because they do not specify how the testers were trained, what the criteria were for the neurological screening or how it was carried out, or how literacy was determined; in addition, cell frequencies are not given for two of the three samples, standard deviations are missing, no data on reliability or any type of validity are given, and there is no rationale for test selection.

Fasteneau and K. M. Adams (1996) reviewed a 1991 norm compilation by Heaton, Grant and Matthews noting that there are too many age–sex–education cells (in itself a welcome change from the frequent failure to account for these variables at all), so that for clinical decisions the requirement that there be a minimum of 30 individuals per cell is not met: In some cells there may be four subjects and in some only one. Because of this and other methodological flaws, they concluded that the compilation has significant limitations.

[2]This searchable database is available by application to the South African Clinical Neuropsychology Association, Institute for Social & Health Sciences, University of South Africa, P.O. Box 392, Pretoria, South Africa.

## THE WAY FORWARD

The way to break the vicious circle of uncertain construct validity, lack of norms, and the resulting ambiguity of psychological assessments is to acknowledge that test scores are the royal road to constructs, and scores have priority over constructs because they are the raw material from which constructs are made. The route to the determination of construct validity thus lies through the assembly of score collections on convergent tests from large samples representative of the target ethnocultural group.

The minimum requirements for adequate score gathering in a diversity of national and cultural settings are summarized here.

### Uniformity of Test Administration

In the first place, this requires a test administration manual at the level of detail of, for example, the World Health Organization's *Operational Guide* (1986) for the WHO–NCTB, the Operational Guide for the Performance Probe Battery (Nell & T. Taylor, 1992), or the administration manual for the *Core Battery of Psychological and Neuropsychological Tests for Persons with Less Than 12 Years of Education* (Nell & Maboea, 1997). For the latter, there are also verbatim instructions for the practice and extended practice loops. These manuals tell examiners in considerable detail how to introduce themselves, how to set clients at their ease, how to lay out the test materials, and how to respond to client puzzlement.

Second, test administrator training must bring novice and experienced administrators to comparability. Of interest in this connection is the procedure followed by the Psychological Corporation. Examiners recruited to standardize the WAIS–III/WMS–III "completed a detailed background questionnaire [and] were very familiar with individual assessment presentation with WAIS–R administration" (Psychological Corporation, 1997, p. 36). Quality control measures included face-to-face or videotape training sessions and the submission of practice protocols, on which detailed written and oral critiques were provided within 48 hours. Each examiner-scored protocol was then rescored by two independent scorers, followed by a discrepancy analysis for each protocol; to prevent scoring drift, anchor protocols were developed and applied (pp. 36–39).

### Demographic Variables

The effect of demographic (or personal) variables on psychological and neuropsychological test scores is dramatic. Heaton, Grant, and Matthews (1991) acknowledged that performance on most neuropsychological tests is strongly related to age and education; on tests of motor speed and strength, gender is important. They continued: "A significant amount of variability in almost all test scores (more than 40% in some cases) could be accounted for by a single demographic variable" (p. 3). These effects multiply in culturally different settings, and have for example been extensively documented in South Africa (Brown et al., 1991; Colvin, Myers, Nell, Rees, & Cronje, 1994; London et al., 1997; Nell & Kirkby, 1997; Nell et al., 1993).

The following are among the most important demographic variables.

***Quality of Education.*** The educational level of a youth who has completed 12 years of education in an underresourced rural school in a developing country cannot be compared with an equivalent period of schooling in Tokyo or Toronto; even within the same country, a given number of years in a violence-plagued inner-city school cannot compare with the same period in a quiet university town. Years of schooling is therefore a crude indicator of educational attainment because it says nothing about those aspects of school quality that are taken for granted in Western settings. Accordingly, as many quality indices as possible should be given. Among these are the level of teacher training and the teacher–student ratio in the school as a whole and in the last class attended by the client, and the availability and quality of school equipment: for example, in primary school, does each child have a desk? Are there sufficient writing instruments and paper for each child? Is there heating and electricity? Is there a school library, and if so, what are its quality and accessibility? Do the children in the higher primary grades have access to computers?

In high school, most of the previous questions also apply, but it is also important to know whether or not there is a science laboratory, and about its quality: Is the instructor capable of giving adequate demonstrations, and is there sufficient hands-on equipment for each pupil to repeat the experiments?

Other important personal variables are *alcohol use* (a replicable and relatively foolproof method of quantifying this is described in London, Nell, Thompson, & Myers, 1998), and detailed *exclusion criteria* for previous head injury, neurological, or psychiatric disease. The *occupational level* of the individual should be specified, detailing any supervisory or managerial functions.

***Environmental Variables.*** A standardized and replicable rating scale (such as an updated version of that developed by Claassen, 1985) should be used to describe the socioeconomic level of the home environment (e.g., whether it has running water and electricity, the number of people per habitable room, and the presence in the home of books, magazines, radio, television, and a home computer).

Similarly, the home and the current setting of the individual should be described (e.g., is this an isolated back country cabin or hut, a village, a town, or a city?) and also that person's position on the rural–urban continuum by giving the number of years of exposure to city life, with its plethora of environmental stimuli.

Finally, to facilitate reanalysis by other researchers, test results should be reported both as standard scores and as raw scores prior to adjustment or conversion.

### Determining Construct Validity

The technical brilliance of factor analysis has tended to blind investigators to the merits of less sophisticated techniques. In the first place, a correlation matrix is closer to the client–test interface than first- or second-order factor analyses, and more transparent to qualitative analysis. If when using a correlation matrix the cutoff for statistical significance is set at a sufficiently stringent level, and nonsignificant values of the correlation coefficient are suppressed, then clusters of variables seem almost to leap out of the matrix to impress themselves on the researcher's attention, and it can readily be determined if hypothetically cognate measures cluster as they ought, or if tests with clear construct validity in the standardization group have taken on a new and unexpected meaning in the minds of the target group.

Second, clinical acumen, which need not only be the last resort of examiners who have not done their homework, has an important role. No clinician recording the responses of a semiliterate client to Controlled Oral Word Generation by letter of the alphabet or Trails B will have any doubt that these are meaningless tests for that individual; the results would be similarly insignificant for the Wechsler Picture Arrangement Subtests in an exotic setting.

However, the most powerful tools for the determination of construct validity remain exploratory and confirmatory factor analysis. Carroll (1993) wrote that exploratory factor analysis "lets the data speak for themselves" (p. 82) and does not thus require that hypotheses be set up in advance. Confirmatory factor analysis on the other hand is best employed to test the factor composition of a set of variables or the structure of factors that have been hypothesized. But, wrote Carroll (1983), it is mandatory that the hypotheses to be tested "have excellent logical or psychological support in some theory of individual differences" or in prior factor analyses of similar data sets. But, if data "can be generated about equally well under several alternative models," it is difficult to choose between models (p. 82).

Fortunately, a readily applicable model is available, namely the nine indices verified for the WAIS–III and the WMS–III by confirmatory factor analysis (Psychological Corporation, 1997). Table 10.2 (chap. 10) shows that the WHO–NCTB, the WAIS–III, and the WMS–III are readily accommodated within the parameters of this model.

On theoretical grounds, three hypotheses can be formulated for fully test-wise subjects: First, exploratory factor analysis will confirm that the additional items in Table 10.2 that are not subtests of the WAIS–III or the WMS–III, but have on prima facie grounds been allocated to the indices for processing speed, perceptual organization, auditory memory, auditory recognition memory, and visual memory, do in fact cluster with the index in question; second, the three items under visuomotor abilities, which is a well-established factor in its own right (Carroll, 1983), cluster together; and finally, on confirmatory factor analysis identical or similar factors to those established for U.S. subjects will emerge.

For non-test-wise subjects, who are the focus of this inquiry, no predictions are made. Test clusters compatible with the factor structure of the Wechsler tests may or may not be confirmed. As factor analytic studies in a variety of settings are carried out, the data will indeed speak for themselves.

## NORMS AS MAGIC NUMBERS

> A person with one watch knows what time it is; a person with two watches is never sure. (Anonymous)

With a test kit in one hand and a norm table in the other, it is easy for inexperienced clinicians to feel diagnostically infallible—in part because of psychology's infatuation with numbers—but not even good norms can be elevated to magical status. The purpose of the interpretive cautions that follow is unkind: It is to take the magic out of norms, and to do so by replacing certainty with uncertainty. Having two watches will better serve client welfare than groundless certainty.

In the first place, a multitude of interpretive cautions apply even to the best of norms. The strictures of Fasteneau and K. M. Adams (1996) on the otherwise excellent 1991 norm compilation by Heaton et al. indicate some of the hidden pitfalls of which the cautious examiner must be aware. Similarly, a series of reanalyses by Matarazzo and his colleagues of data from the WAIS–R standardization sample (Matarazzo, 1990; Matarazzo & Herman, 1984, 1985; Matarazzo & Prifitera, 1989) show that significant cautions apply even to the Wechsler tests, which are the benchmark instruments of psychological assessment. No similar reanalyses of the WAIS–III standardization data are yet available, so that the generalizability of the material that follows is unclear. But because the Wechsler scales are increasingly being translated and renormed for use in culturally different settings, it is nonetheless useful to present a summary of the main findings of this series of papers.

### Test–Retest Variability

On the WAIS–R and the WAIS, individual test–retest variability is high: Of 119 subjects who were retested on the WAIS–R 5 to 7 weeks after the initial testing, 38% gained between 5 and 15 scaled points on the Verbal IQ score, and 50% gained the same amount on the Performance scale. On individual subtests, 11 of the 119 subjects gained a full standard deviation on Arithmetic, and 21 did so on Object Assembly (Matarazzo & Herman, 1984). One cannot therefore say with certainty on the basis of any one testing on the WAIS–R or similar tests that the obtained score represents the client's true level of ability. Moreover, within these limits, the supposed "immutability" of IQ scores is shown to be fallacious.

*The standard error of measurement* (SEM) of individual subtests on the WAIS–R is substantial. On the Object Assembly subtest, for example, it is 1.54 scaled points, half a standard deviation. To give a score at a 95% confidence level, one needs to take one SEM above and below the obtained score (Matarazzo, 1990).

### Subtest Scatter

It is still widely believed that intelligence, like water, has a single level, flowing so as to obliterate hills and valleys. On the contrary, "subtest scatter is characteristic of normal subjects" (Matarazzo, 1990, p. 1008): Among the 1,880 subjects in the WAIS–R standardization sample, the mean difference between highest and lowest subtest scores was 4.67 scaled points on the Verbal scale and 4.71 on the Performance scale (differences that are significant at the 5% level of probability!). Further, 69% had 6 or more scale points difference between their highest and lowest subtest scores, which is a difference of two standard deviations.

### VIQ–PIQ Differences

Wechsler (1958) proposed that a difference of 15 scale points or more between the VIQ and PIQ scaled scores was diagnostic of brain damage. This is incorrect both for the WAIS–R (Matarazzo, 1990) and for the South African Wechsler (Murdoch, 1982). Although the mean VIQ–PIQ difference in the full WAIS–R standardization sample is zero, the standard deviation is 11.12, with a range from −43 to +49 scaled IQ points. This in effect means that "even very marked VIQ–PIQ scatter is not necessarily pathognomic when it constitutes the only evidence of impairment" (Matarazzo, 1990, p. 1010). Moreover, in a sample of 365 subjects with no history of neurological or psychiatric illness, Bornstein (1985) showed that as many had a performance IQ in excess of their verbal IQ as those who scored more on the verbal than on the performance scale. Bornstein (1986a) amplified on these data, noting that the statistical prediction is that a difference of 10 points between verbal and performance scale scores is required for significance at the .05 level; however, empirical observation of the actual occurrence of discrepancy scores showed that a difference of 23 points (1½ standard deviations) between the two scales was required in order to attain significance.

### Cutting Scores

A further set of cautions applies to the Halstead–Reitan Neuropsychological Test Battery in relation to the published (and still sometimes used) cutting scores (Russell, Neuringer, & Goldstein, 1970). Bornstein (1985, 1986a, 1986b) has demonstrated the risk of false positives in using these cutting scores to distinguish between normal and brain-injured persons. Of the entire sample of 365 healthy subjects, 80% were classified in the impaired range on finger tapping (and all males between age 40 and 69 without a high school education!), 62% as impaired on the Grooved Pegboard Test, and 20% on Part B of the Trail Making Test (Bornstein, 1986a, p. 419).

### Score Distribution

Mean scores do not take score distribution into account. Thus, the percentage of individual scores in the impaired range on the Trail Making Test in Bornstein's sample (1985), although the mean was normal, rose to 40% for 4 of 12 subgroups, and to 75% for 1 subgroup (see also Bornstein 1986b; Finlayson, Johnson, & Reitan, 1977).

Bornstein's (1986b) final comment applies to norms in general, and deserves to be inscribed on consulting room walls in letters big enough to be read by the light of the new moon:

> The implications of increased false positive errors, particularly when reliance is placed on "expected proportions" or "rules of thumb," should be an important message to those involved in use of these measure. . . . These data are in dramatic contrast to the practices of those who base their interpretations [on] readings or attendance at brief workshops. [They] serve to underscore the remarkable variability in human behavior, and . . . the necessity for neuropsychologists to develop large scale normative data bases, upon which basic principles and patterns of test performance can be established. (p. 19)

Let me end this chapter with a cautionary tale that illustrates the weight of this warning.

The Van and its Cargo

While I was working on this chapter, a client's compensation claim was about to come to trial. The trial lawyer, a man with a formidable courtroom reputation, was reviewing the neuropsychological evidence with me. With us were his legal assistant, a clerk from the attorneys' office, and an industrial psychologist. Leaving my own waning abilities out of account, an electrical hookup of the brains in that room would have skittered the pens of an EEG machine off the page. Matters proceeded amiably enough over a round of toasted sandwiches and coffee until we got to the part of my report that recorded the client's long struggle with a multistep arithmetic problem. "What's this problem he had so much trouble with?" asked the trial lawyer. I pulled out my copy of Walsh's *Understanding Brain Damage*, and read aloud: "A van and its cargo are worth $16,000. The van is worth $700 less than the cargo. How much is the cargo worth?" (1985, p. 238).

"Why don't you try it yourselves?" I suggested generously, implying, as Walsh does, that reasonably intelligent people with intact brains would have little difficulty in solving this puzzle. Silence descended on the room. The lawyer, the assistant, the clerk, and the psychologist lowered their heads, took up their pens, and began scribbling figures and crossing them out, as I have often watched clients do. Within the first minute, there was a flurry of wrong answers, $8,700 being the most popular. There was more silence, and more scribbling. When 5 minutes had passed, the lawyer lifted his head, tossed his gold pen onto the table, and asked with asperity, "And what, if I may ask, is the meaning of this humiliation you have inflicted on us?"

What, indeed?

Here were three intelligent and successful professionals. None was a mathematician, but only clear thinking—and not mathematics—was required to solve the problem. Yet, exactly like my brain-damaged client and many others like him, they had struggled for 5 minutes and come up with exactly the answer he had offered, $8,700, giving a difference of $1,400 between van and cargo. Like the client, they had not divided $700 by two—the insight I had so readily accused him of lacking—so that the difference between the greater and the lesser figure would be $700, not $1,400.

Writing about the impact of what she called "wretched childhoods" on later development, Richter (1990) remarked that it is "not difficult to find incidents of misfortune in the backgrounds of people who later developed problems." She went on to cite Sigmund Freud, himself a great master of hindsight, who wrote as follows:

> So long as we trace a development from its final outcome backwards, the chain of events appears continuous and we feel we have gained

*(Cont'd. on next page)*

*(Box cont'd. from previous page)*

> an insight which is completely satisfactory and even exhaustive. But if we proceed the reverse way, if we . . . try to follow these [premises] up to the final result, then we no longer get the impression of an inevitable sequence of events which could not have been otherwise determined. (in Richter, 1990)

Neuropsychologists must learn the lesson that Freud taught himself: Predicting backward to a traumatic episode, whether psychic or a bang on the head, is easy and invariably correct. But diagnosis is also prognosis, the art of predicting forward so as to extrapolate from the consulting room to real life. That is what norms are for: to ensure that if attributions of deficit are made, they are founded not on the psychologist's personal certainty that all normal people should get the item correct (and this is the assumption implicit in the presence of a test item in a book about brain damage), but on hard data that show that most people in the target population do in fact get the item right: that is the meaning of the humiliation in the lawyer's chambers.

Chapter 6

# A Behavioral Frame: Neuropsychology as a Transferable Technology

The professional core of clinical neuropsychology is its highly differentiated ability to describe the behavior of brain-damaged persons. However, the reader of current neuropsychology handbooks and journals is likely to arrive at two incorrect conclusions about the nature of current neuropsychological practice.

## THE PREPONDERANCE OF DIFFUSE INJURIES

The first error is to believe that most of the cases brought to a neuropsychologist's attention are discrete focal injuries, and can be described in terms such as "infarct of the inferior left angular gyrus" or "left frontal orbitobasal contusion." This is certainly true for neuropsychologists in tertiary acute care settings. But the picture is very different for forensic, rehabilitation, and private practice neuropsychologists who work outside neurology-neurosurgery settings, for isolated psychologists, and for most psychologists in the developing countries. The clinical consequence of the epidemiology of brain damage (reviewed in greater detail at the beginning of chap. 7) is that the overwhelming majority of individuals requiring neuropsychological diagnosis and treatment are the survivors of motor vehicle accidents, assaults, and falls. The violent forces that these events unleash in the brain cause extensive diffuse damage that cannot be accommodated within the correlational neuroanatomy of classical neuropsychology. These are not discrete focal injuries that give rise to such marvelously bizarre syndromes as unilateral neglect, pure alexia, or mistaking one's wife for a hat. Higher intellectual and personality functions are diffusely disrupted, and the neuropsychologist's task becomes to describe the ways in which intellect and behavior have changed rather than to attribute these changes to specific brain locations.

## ASSESSMENT VERSUS TESTING

The second error is to believe that "neuropsychology" and "testing" are synonyms, and to be a competent neuropsychologist means that one is an expert in the application and interpretation of "neuropsychological tests." On the contrary, what neuropsychologists do best is describe behavior—as it was in the past, as it is now, and as the individual is likely to behave in the future. To do so, they always draw on the same two sources of information: what they are told by the client and the client's circle, and what the tests tell. The tests are always interpreted in the light of the interview information, and very seldom the other way around.[1] In other words, the essence of the neuropsychological specialization is the systematic description of behavior, based on descriptions of real-life behavior, and on tests, which are simulations of real life (for an excellent statement of this position, see Matarazzo, 1990).

A third error lies latent in this notion of "neuropsychological tests," because there is no theoretical basis for distinguishing between neuropsychological tests and the generic class of "psychological tests." The tests neuropsychologists customarily use are not uniquely neuropsychological. Some (such as the Wechsler tests) are used for all psychological assessment, and others could with benefit be more widely used by generalist clinicians. Indeed, the assessment of perceptual-motor integrity, memory in the verbal and nonverbal domains, and higher level problem solving should become part of every comprehensive psychological assessment, not locked away in a cupboard marked "Neuropsychologists Only." This applies equally to classic neuropsychological tests such as Trail Making and the Wisconsin Card Sorting Test, and more recent additions to the repertoire like the Austin Maze and the Tower of London.

## NEUROPSYCHOLOGY AS A DESCRIPTIVE DISCIPLINE

In the context of diffuse axonal injury, neuropsychology is thus a descriptive discipline, and this descriptive impetus has been supported by the cognitive

[1]The tests take precedence over complaints only if the client and the informants seem consciously or unconsciously to be exaggerating their difficulties.

revolution in psychology. Karl Lashley's address to the Hickson Symposium at the California Institute of Technology in 1948 marks the beginning of the cognitive revolution in psychology, writes Gardner (1987). In "The Problem of Serial Order in Behavior," Lashley argued that activities such as playing the piano and speaking could not be conceptualized within a stimulus–response paradigm; on the contrary, it is necessary to accept that a neural control system existed to regulate this rapid serial unfolding of complex behavior. Soon after, Noam Chomsky (1959/1964), then a young linguist at the Massachusetts Institute of Technology, struck a cognitive blow at Skinner's behaviorist account of language from which radical behaviorism would find it difficult to recover (Gardner, 1987, p. 192). Suddenly, it was again respectable in psychology to talk openly about language, logic, and reasoning, and the scramble to find purely behavioral evidence of the psyche's operations, and tease out hard evidence of the neural substrates of cognitive activity, was over at least for a while.

What is the outcome of these two trends—that most peacetime cases of brain damage that come to the attention of neuropsychologists are diffuse axonal injuries, and that it is not only respectable but also desirable to describe the effects of brain injury in cognitive terms? As a result of these trends, clinical neuropsychology has come to lay increasing emphasis on the analysis and description of behavior, and relatively less on the correlation of disrupted functions with discrete loci of cerebral malfunction. The language neuropsychologists speak when they are dealing with case material is the language of behavior rather than of neurology. Here, for example, is a case description from Luria's *Higher Cortical Functions in Man* (1962/1980):

> A patient with a severe traumatic lesion of both frontal lobes, followed by the development of a cyst which replaced the brain tissue in the frontal region, when asked to draw a square, drew three squares, thereby filling the top line of the paper. When asked to draw one square, he drew lines all around the edges of the paper. At this time the doctors were talking to each other a little way off. The patient picked up the word "pact" from their conversation and thereupon wrote the words "Act No." When one doctor murmured to the other, "It is rather like the experiments with animals, after extirpation of the frontal lobes," the patient picked up the word "animal" ("zhivotnoe" in Russian) and wrote down the words "o zhivotnovodstve" (meaning "about stock breeding"). When the doctor asked the nurse, in a whisper, the patient's name, the patient immediately wrote it down—"Ermolov." Thus, the patient's whole behavior was determined not by specialized, selective systems of connections, but by irrelevant stimuli, which he perceived without any kind of discrimination and which quickly evoked a motor reaction. (p. 306)

Another neuropsychologist with a highly differentiated knowledge of the central nervous system is Kevin Walsh. This is one of his case descriptions:

> This 23-year-old nurse suffered an open compressed skull fracture with tearing of the dura and damage to the left frontal lobe when she was thrown onto the road when the touring bus in which she was travelling in Europe was involved in a collision. She was deeply unconscious when admitted to a German hospital. . . . At four months she was repatriated to Australia. On admission to our hospital there was gross impairment of language, memory and frontal lobe function. Over the next three months she regained motor function and when first examined neuropsychologically she was continent and could walk with a stick. Speech therapy had been commenced and the state of her language functions was monitored because of the known left frontal damage though more widespread damage was probably present. . . . She understood the significance of logico-grammatical constructions. However, independent meaningful expression was disturbed. Verbal reasoning was extremely concrete. . . . There was a childish character to many of her replies with disinhibition and attempts at simple jokes.
>
> What is an armadillo? "An arm with a dillo on it." Why do we brush our teeth? "Keep them clean, white as normal. Keeps them happy. Oh, that was silly." . . . Her lack of selectivity probably due to lack of inhibitory control was seen on the Verbal Fluency test. . . .
>
> S. "skin, scoot, school, skin (I said skin), skittle, sun, son, sound, skin, skittle, son." (Walsh, 1985, pp. 179–181)

It is interesting that both the Luria and the Walsh case descriptions begin in the focal, locationist tradition. We are told in the first case of a "severe traumatic lesion of both frontal lobes," followed by the development of an *ex vacuo* cyst, and in the second of "an open compressed skull fracture," "tearing of the dura," and "damage to the left frontal lobe." But as the case histories unfold, these details recede from foreground to background. The Walsh case is reclassified from focal to diffuse ("more widespread damage was probably present"—hardly surprising given the violent forces exerted when this woman was flung from a fast-moving vehicle), and the concluding paragraph consolidates the behavioral framework that has defined this case study:

> A year later the situation was unchanged though she had now moved to a centre where she continued an interest in art and craft classes. She was described as warm and friendly with only partial insight into her problems with doubts of her own worth. She lived at home with her father and moved around freely on public transport. Further attempts to employ her even on a voluntary basis had failed. To many lay observers she still appeared to be a "normal" young woman. The eventual loss of her father will bring about a serious crisis in her life since she is still unable to manage the planning of her life's activities. (p. 182)

## CAUSALITY IN MEDICINE AND IN PSYCHOLOGY

There is a difficulty with the argument in the previous section. Under the impact of the natural sciences, we have come to believe that to understand a phenomenon, and certainly to control or change it, it is not enough to describe what one sees. Its cause must also be known. Foucault (1963/1989) wrote that in medicine, a profound change, hinging on precisely this point, came about at the close of the 18th century. Until then, disease had been described in terms of its essence. Phenomena with a similar observable effect, that is, with the same symptoms, were thought to have a common cause. For example, paralysis, apoplexy (stroke), and syncope (fainting) all result in the abolition of voluntary movement and were thus held to be manifestations of the same disease: "When they become dense enough, these similarities cross the threshold of mere kinship and accede to unity of essence" (1963/1989, p. 7). But from the 19th century, what Foucault called "the medical gaze" has quite precisely inverted this belief, substituting "the absolute values of the visible . . . [for] systems and all their chimeras" (p. xii). Modern medicine thus knows that the same symptoms (e.g., fever, muscle pain, weakness, and lethargy) may be produced by a great variety of diseases (e.g., a common cold, influenza, malaria, or measles). Each disease is in turn linked not to its symptoms, but to its cause—a precisely specified and unique pathogen.

Accordingly, a neurologist who wishes to make a diagnosis that determines treatment and prognosis may not offer a description of behavior and stop short there: "The client is confused, incontinent, and walks with a wide-based gait . . ." Is this poor fellow suffering from mercury poisoning, multi-infarct dementia (MID), a degenerative disease, a tumor, or normal pressure hydrocephalus (NPH)? For each, treatment and prognosis differ. In medicine, the cause is of paramount importance.

In psychology, this 19th-century inversion has not taken place. A single psychic "pathogen," such as the loss of an attachment figure in early childhood, may give rise to lifelong underachievement or high creativity (Ochse, 1990), to depression or the energized fanaticism of the True Believer. The epidemiological evidence shows "clear correlations between most forms of neuropathology and a small group of causes,"[2] wrote Albee (1986), who contrasted "real diseases" with psychological distress. Metaphorically, psychology and its sister discipline, psychiatry, continue to operate in Foucault's premodern era of essences related to symptoms, which in treatment have precedence over cause.

---

[2]These are "emotionally damaging infant and childhood experiences; poverty and degrading life experiences; powerlessness and low self-esteem; and loneliness, social isolation, and social marginality" (Albee, 1986, p. 891).

Neuropsychology functions in two domains: within neurology as a medical causality-specific science and within psychology as a discipline focused on behavioral outcome rather than cause. In the neurological domain, the neuropsychologist functions attributively, using psychological instruments to determine, for example, if a dementing condition arises from NPH, MID, or a degenerative process. But in the psychological domain, the neuropsychologist's diagnostic expertise is applied not to the specification of a cause but to the description of its effects. Many classes of central nervous system insult, and especially diffuse closed head injuries, have the character of a psychic pathogen, and, like early childhood loss, give rise to a multitude of symptoms. For the neuropsychologist, it is describing and understanding the behavioral consequences of an injury that are important, rather than a specification of the injury itself, thus reversing the relative importance of pathogen and symptom in the medical diagnostic system.

As we shall see throughout Part III, this shift in neuropsychology from the neurological-locationist tradition to a behavioral-descriptive frame has changed many aspects of Western neuropsychological practice, and at the same time has legitimated the practice of a behavioral neuropsychology with culturally different clients.

## NEUROPSYCHOLOGY AS A TRANSFERABLE TECHNOLOGY

This redefinition of neuropsychology as a cognitive science has major implications for neuropsychological services in isolated and developing country settings. Holtzman, Evans, Kennedy, and Iscoe (1987) defined a "transferable technology" as one that can effectively be devolved by fully trained personnel to those with a lesser level of training. To the extent that neuropsychological assessment is conceptualized as a branch of the neurosciences, it remains inaccessible to most psychologists. If, however, the neuropsychological assessment of clients in community settings, as distinct from hospital inpatients, is redefined as the componential analysis of behavior within a cognitive science framework, behavioral neuropsychology is mainstreamed with other forms of psychological assessment, and becomes accessible to psychologists and other health workers with an adequate background in cognitive psychology. Such practitioners are thus enabled to offer services at a level of behavioral analysis and description that is useful to the client and the family, and serves as a guide to intervention and therapy.

The need is certainly there. There are many brain-injured people in the developing countries of Africa, Asia, and South America, as well as those who work in isolated areas of the developed countries, to whom specialized clinical neuropsychology services are unavailable or inaccessible. The question is whether basic skills in behavioral neuropsychology can be devolved, and whether this would be professionally desirable.

## HUMAN WELFARE AND GUILD CONCERNS

There is an inherent tension in all health care professions between human welfare and guild concerns (Nell, 1989). Thus, making psychological skills more widely available may undermine the status and income level of expert practitioners. Glossing over this tension or ignoring the professional interests of practitioners such as physicians, registered nurses, or clinical neuropsychologists, is counterproductive. The provision of sophisticated health care requires long and expensive training, which can only be provided and made attractive to practitioners if their professional interests are protected. If not, there will be no specialists with skills to devolve. Blurring the distinction between specialist neuropsychologists and the level of behavioral description and diagnosis addressed by this book will not be helpful either to health care and the clients of the health care system or to professional development.

A useful model for a limiting description of levels of practice is provided by Moses, Golden, Ariel, and Gustavson (1983, pp. 41–44), who described four increasingly complex levels of neuropsychological practice. In this scheme, practitioners may be designated as fully qualified clinical neuropsychologists only at the fourth level, which requires that they be able to describe on the basis of interview and test data how the brain of a client is functioning, attributing apparently disparate symptoms to a common underlying cause in central nervous system pathology. Such localization can be a means of generating further hypotheses and generalizations about functioning in related but dissimilar circumstances.

I have adapted their scheme as a three-level description of practice appropriate for the definition of behavioral neuropsychology.

*Level 1* is screening for brain impairment, using one of the many available approaches (Berg, Franzen, & Wedding, 1987; Nell, 1985). However, the question will very seldom be, "Does this individual have brain impairment?" but rather "Does the documented brain impairment in this individual affect his or her thinking and behavior?" This simple "yes" or "no" question can be answered by psychologists who have had introductory courses in the principle manifestations of brain impairment (chap. 7), and in behavioral interviewing (chap. 8).

*Level 2* interpretation requires that the practitioner describe the client's thinking and behavior in terms of an hierarchical sequence of intellectual and behavioral functioning. Such parameters are more fully set out in Part III of this book. For intellect, these are an adequate level of arousal and the integrity of perception, of complex sensory integration, of memory, and of the planning and regulation of intellectual activity. The hierarchical organization of personality functioning relates to the adequate inhibition of impulse-governed behavior, the maintenance of social appropriateness by the correct perception of social cues, of conversational pragmatics, and, at the highest level, of satisfactory intimate relationships with both family of origin (parents and siblings) and the marital family (spouse and children).

From this detailed description of the ways in which behavior has or has not been disrupted by brain impairment, inferences can be drawn about functioning in related areas, such as school or university, in the workplace, and in social networks.

*Level 3* interpretation attributes specified changes in intellect and personality to their correct underlying origins in brain impairment. Thus, apparently disparate symptoms—such as hyperverbality, increased irritability, and confusion in doing jobs around the house or at work—can all be related to a common underlying cause in central nervous system pathology. The cause must be closely specified. It is not good enough to use gross or pseudo-psychiatric terms such as "organic brain syndrome" or "frontal lobe syndrome." The hallmark of full specialization is the ability to make fine-grained causal attributions that derive from detailed knowledge of central nervous system control of thinking and behavior, and the mechanisms through which such control is exercised. The specification of these relations allows for strong predictions of performance in other task and social areas. It is only at this level that the designation "clinical neuropsychologist" can be used.

## THE ISOLATED PSYCHOLOGIST

Just as country doctors are compelled by human welfare concerns to carry out a range of complex medical and surgical interventions that their big-city colleagues would summarily refer to specialists, so are psychologists in isolated settings constrained by the same human welfare concerns to diagnose and treat individuals that, given a kinder reality, they would have preferred to refer to a specialist, whether in learning disabilities, sexual dysfunction, depression, or brain damage.

In these circumstances, a well-defined behavioral neuropsychology will enable its practitioners to formulate a fine-grained componential picture of the behavioral and intellectual processes that have been disrupted by brain damage, and to do so by drawing on and amplifying the knowledge base all psychologists share in the cognitive sciences. This behavioral specification can be made without reference to the highly differentiated knowledge of human central nervous system anatomy and pathology that characterizes the specialist clinical neuropsychologist.

Devolution of skills in clinical neuropsychology thus means devolution from Level 3 to Level 2 practitioners—a behavioral-descriptive, nonlocalizing neuropsychology—and the establishment of Level 2 practice as an essential health care modality. Moreover, by using this book as a teaching manual, psychologists, physicians, occupational therapists, and other conceptually trained health workers will be able to provide a carefully circumscribed set of skills to primary health care workers such as clinic nurses and village health workers, and also to mental health personnel in both primary care and referral hospital settings. This is the material contained in Part III, namely, a simple and relatively concrete system for identification of the concussion syndrome and for differentiating it from mental illness. This may appear to be a trivial contribution to health care, but on the contrary it is substantial both for its reassurance value to victims of brain damage and their families (explaining that the damage will not "spread" and is not progressive) and because it avoids needlessly hospitalizing or overmedicating brain-damaged individuals who constitute no danger to themselves or others.

In this way, the needs of the isolated psychologist are addressed, and at the same time a way to develop a wider reach of nonspecialist neuropsychological services is opened.

*Part* III

# THE PRACTICE OF CROSS-CULTURAL ASSESSMENT

Chapter 7

# The Cardinal Manifestations of Traumatic Brain Injury

This chapter and the next set out a method of eliciting as complete and well-rounded a picture as possible of the client's behavior. To do so, it is necessary to have a clear view of the *behavioral* changes that follow most traumatic brain injuries, and a method of eliciting descriptions of these changes. These matters are dealt with in sequence in this chapter and the next, which are both about the diagnosis that comes *before* testing. This "before" has a twofold primacy: First, behavior in the real world has precedence over behavior in the test room, and second, it is the client's behavior that determines the selection of tests, their prognostic interpretation, and the structure of the neuropsychological report (chap. 11).

* * *

Worldwide, diffuse traumatic brain injuries are the most commonly encountered type of brain damage. These are caused by the application of violent accelerative or decelerative forces to the head: *Accelerative* forces arise in pedestrian motor vehicle accidents and from blows to the head, and *decelerative* forces occur during heavy falls, and for the occupants of motor vehicles involved in accidents. In the United States, traffic accidents account for between 60% and 80% of all brain injuries (Romer et al., 1995). In Bangalore, India, 62% of brain injuries are caused by traffic accidents (Channabasavanna, Gururaj, Das, & Kaliaperumal, 1994). In South Africa, traffic accidents cause 73% of all brain injuries among White males, but cause only 30% of such injuries among Black South Africans.

This is the result of the very high rate of interpersonal violence in this second group, which accounts for 51% of all brain injuries: Nearly all these injuries are caused by blows with blunt weapons, again causing diffuse injuries (Brown & Nell, 1991; Nell & Brown, 1991). In the developing countries, pedestrians, and especially children, are at specially high risk. Many roads are unkerbed and without sidewalks, forcing pedestrians into the traffic flow; because law enforcement is poor, traffic travels at unsafe speeds, ignoring traffic lights and stop signs. Prevention in these settings is extraordinarily difficult because civil society is too weak to exert effective advocacy pressure for better—and less corrupt—policing.

At the same time, the lessons of the Vietnam and Middle East wars—that field stabilization and rapid evacuation of head injuries dramatically improves outcome—have revolutionized peacetime evacuation and treatment techniques. In most big cities, paramedics are trained to stabilize accident victims on-site and provide life support en route to the emergency room. These methods combined with sophisticated intensive care technology insure the survival of an ever-increasing number of severely injured head trauma cases (especially road accident victims), who in earlier times would have died at the scene of the injury or soon afterward, and therefore increase the prevalence of diffuse brain injury in the population.

This very large pool of diffuse traumatic injuries is augmented by other etiologies of diffuse brain damage (i.e., anoxic episodes, neurotoxicity, HIV infection, and degenerative processes such as Alzheimer's disease and Parkinsonism). But unlike traumatic brain injury, the pathogenic degenerative conditions may have changing focal elements at different times in the course of their development and present problems that are similar but not identical to the usual sequelae of traumatic closed head injuries.

Two matters relating to the focus of this book arise. For the epidemiological reasons given earlier, the focus is throughout on *traumatic* brain injury, which constitutes the overwhelming weight of most neuropsychologists' case load; but, *mutatis mutandis,* the examination and assessment techniques described in this and the following chapters are applicable to all diffuse injuries. Second, and again for epidemiological reasons, the focus should be on mild and very mild brain injury, because such injuries, with an admitting Glasgow Coma Scale of 13 to 15, constitute some 80% of all live hospital admissions (Romer et al., 1995). However, this is a difficult focus to maintain: The behavioral manifestations of mild and of severe brain injury differ in degree rather than kind, so that the behavioral manifestations of severe brain injury are subtly or more obviously present in mild and very mild injuries as well. As Walsh (1985) pointed out, teaching proceeds by the method of extreme cases, and most of the case material that follows relates to severe injuries.

## THE FOUR COMPONENTS OF DIFFUSE BRAIN DAMAGE

Behavior changes in predictable ways after diffuse brain damage. This chapter systematically describes these changes under four main heads: changes in *arousal*, in *personality*, in *thinking*, and in *physical function*.[1] But all diffuse brain injuries result in a pattern of changes that is a seamless whole. For example, each alteration in arousal level prefigures a change in personality and a change in thinking; these linkages in turn suggest what the underlying neural substrates might be.[2] This seamlessness has two implications: First, the headings that follow are pointers, not boundaries. Second, for the experienced neuropsychologist, it means that as each symptom is elicited, it triggers a clinical hypothesis (or a series of hypotheses) about likely related changes. For example, if the caregiver says, "He doesn't seem to notice things any more. If the food's burnt or there's no sugar in his coffee, he doesn't say a thing," then the neuropsychologist surmises that the problem is not one of inattention, but rather the loss of gustation, which is usually paired with a degree of anosmia; if the client is anosmic, then the next hypothesis is that he has very likely sustained significant orbitobasal damage and is likely to be disinhibited with a range of related manifestations, from tactlessness to dyscontrol. If so, what is his arousal level, and how does this affect concentration and the problem-solving capacities related to concentration? In this way, in linked loops of hypothesizing that derive from linkages within the nervous system, the skilled neuropsychologist sets about building a diagnostic picture.

The two most pernicious myths about traumatic brain injury are that concussion has no permanent effects (see next section) and that a direct blow to the head is a prerequisite for brain injury. These truisms live on in many standard references: "Unless the head is struck, the brain suffers no injury" (R. A. Adams & Victor, 1989, p. 693). There is, on the contrary, conclusive evidence that significant brain injury can be sustained purely as a result of violent rotational acceleration and without head strike (e.g.,

---

[1]These symptoms co-occur with such regularity that they have the status of a syndrome. Some writers continue to refer to the "postconcussion syndrome," which is as illogical as referring to the prolonged pain after rib fracture as "postfracture pain" (Rutherford, 1989). Fracture says it all, and so does concussion syndrome—or "concussion clinic," a happily nonthreatening term adopted for our unit's outpatient diagnostic and counseling service in Johannesburg.

[2]The pathophysiology of traumatic brain injury is beyond the scope of this book. Interested readers should consult the many excellent reviews of this material (Pang, 1985; Povlishock & Jenkins, 1995)—although as clarity requires, reference is made to the neurological substrates of some of the behaviors described.

by motor vehicle occupants restrained by seatbelts or airbags, and in rear-end collisions giving rise to whiplash without headstrike; Grimm, Hemenway, Lebray, & Black, 1989).

## 1. Hypoarousal

The unchanging hallmark of the concussion syndrome is *hypoarousal*, giving rise to increased fatigue with daytime naps and increased night sleep, fluctuating attention with reduced concentration span, and reduced alcohol tolerance. These changes originate in the ascending reticular activating system of the brainstem, which maintains the brain's level of arousal and therefore also consciousness. Consciousness is abolished by reducing the level of activity of the reticular activating system, as in sleep or by the application of violent force to the head. A useful metaphor that helps clients understand the function of the reticular activating system is "the brain's power station," providing the constantly changing energy supply needed for different levels of cognitive activity. Until the 1970s, it was believed that concussion was a transient and fully reversible condition that caused no lasting damage. Povlishock (1996; Povlishock & Jenkins, 1995) conclusively showed that even mild concussive episodes (GCS 13–15) give rise to diffuse axonal injury (DAI) scattered throughout the brain and brainstem. Although recovery from a single mild episode of concussion will typically be complete, Blumbergs et al. (1995) showed that DAI is present in 75% of the brain sectors of mild traumatic brain injury (TBI) cases at postmortem. However, the effects of even one or a series of mild concussions is cumulative. World champion boxers, who typically give more punishment than they receive, may have *dementia pugilistica* (punch-drunk encephalopathy) by the end of their careers.[3] Long participation in other sports, especially at the professional level, also has its dangers (Maddocks, Saling, & Dicker, 1995; McCrory & Berkowic, 1998; but Tysvaer, Storli, & Bachen, 1989, dispute the persistence of residual effects).

Over and above the fatigue and more important by far are the changes that arise from *hypofrontality*, or reduced activity of the prefrontal areas. This is an invariant consequence of hypoarousal because the frontotemporal areas receive relatively less innervation from the ascending reticular activating system than the retrorolandic areas. The devastating consequences of hypofrontality for personality and memory are considered below.

### *Arousal, the Test Room, and the Real World*

When endogenous arousal is compromised by damage to the reticular activating system, the individual becomes dependent on exogenous arousal

---

[3]Joe Louis, the "Brown Bomber," ended his career as an intellectually depleted greeter in the lobby of a Las Vegas hotel; Muhammad Ali developed pugilistic Parkinsonism.

provided by the environment. The routine, the familiar, and the everyday allow arousal to drop: As a result, arousal is lowest at home with the family, in the classroom, or at work with well-known people. Under these conditions, incidental learning will be poor, as will memory and inhibitory control of behavior. Here in a nutshell is the explanation for a troubling phenomenon: Behavior is always worst at home with the people who care most for the client and are least able to protect themselves (i.e., spouse, children, and pets). In colonial settings, racism may be given unbridled expression, with the worst excesses of rage directed against domestic servants and the local ethnic underclass.

***"Where Is This Monster?"*** However, meeting a new professional in an unfamiliar consulting room, finding oneself the center of attention and responding to all manner of interesting questions about oneself and one's history, is a nonroutine and exciting experience with a high exogenous arousal value. This gives rise to puzzling consequences: During the diagnostic interview, when arousal is high, both temper control and concentration are better than at home. Inexperienced professionals, seeing before them an alert and well-controlled client with evidently good recall, are puzzled at the family's description of an evil-tempered and distractible person, and will ask, overtly or by implication, "Where is this monster you are talking about?"

Matters are further complicated because a hypoaroused client's performance on novel psychological tests, with an interested examiner giving constant encouragement, is likely to be strikingly different to that same person's level of achievement in the classroom and at work. Again, the difference is arousal-dependant: Clearly, the test results of hypoaroused individuals must be interpreted with great caution.

How can this testroom arousal effect be attenuated without detracting from the empathic, hands-on approach advocated in chapter 9? Ensuring that instructions are understood and that there are sufficient practice opportunities is one thing; becoming the client's frontal lobes is another. This will indeed happen if the examiner becomes overinvested in the client's performance and excessively vigilant, for example, by responding even if only by a nod to correct answers and with encouragement to errors: Each such response is an arousal jag (this is Daniel Berlyne's, 1974, evocative term). What needs to happen is that once the instruction phase is over, the examiner withdraws from the field, becoming ground rather than figure.

For example, the Performance Probe Battery described in chapter 9 includes several procedures requiring sustained attention. Initially, we programmed these to sound an alarm if response latencies were excessive, or if three consecutive errors occurred. The examiner would then explore the difficulty with the client, and reset the program to continue. It soon became clear that the alarm and the examiner intervention provided an arousal jag that undercut the validity of the attention measure, and this loop was eliminated from the program.

The backgrounding of the examiner is especially important on tests, such as the Austin Maze (Walsh, 1985, pp. 235–237) or the Wisconsin Card Sorting Test (WCST), that make high demands for sustained attention. For example, on the Austin Maze, as soon as it is clear that clients understand what is required, I show them how to note down their own scores. Then I push my chair back, give every appearance of having lost interest in the test, and often leave the test room; for the same reason, on the WCST, I deliberately make my responses as monotonous and invariant as possible.

### *Examining for Hypoarousal*

To gauge the severity of hypoarousal, ask what time the client would typically go to bed pre-accident, and at what time now; whether it is now more difficult to wake up in the morning; whether there is a diurnal fatigue cycle, getting worse in the afternoon; and whether the client naps during the day if there is an opportunity to do so.

The first victim of hypoarousal is attention, which is by definition effortful (Kahneman, 1973); consequently, so is concentration, which is nothing more than sustained attention. Both become difficult for the hypoaroused person. Ask how long clients typically watch television, how long they read, and, if they are scholars or students, how long they can sit at their desk before needing to take a break. However, answers to these questions have an amplified meaning if frontal distractibility (see "Cognitive Changes") is also present.

### *Reduced Alcohol Tolerance*

Alcohol is a central nervous system (CNS) depressant. In persons who already have a reduced arousal level because of brainstem damage, the additive effects of this additional powerful depressant are immediate and dramatic. In cultures in which heavy social drinking is the norm, individuals who could previously down six or eight beers without becoming visibly tipsy now complain that they feel drunk or nauseous after one or two beers, or that alcohol tastes different. Others give up drinking entirely, but will attribute this change to a principled decision that in their view is unrelated to physical intolerance. If a hypoaroused person continues drinking the same amount as before, the stage is set for major problems, among them rage dyscontrol (Nell, 1990b, 1990c) with the risk of injury to self or others, arrest, and ongoing involvement with the criminal justice system. In examining for alcohol tolerance, the basic questions to ask concern how many beers, glasses of wine, or shots of hard spirits clients would typically have at a sitting before, and how many they have now.

***Noise Intolerance.*** The human auditory system has remarkable selective powers. If a conversation that interests a person is taking place at the other end of a crowded room, after a few minutes that person can "tune in" despite the masking noise all around: This party trick (it is sometimes called the cocktail party effect) needs synergistic interaction between the frontal inhibitory centers that mediate orienting responses and the receptive language systems. Hypoaroused individuals are generally more distractible, responding to stimuli that are irrelevant to the task at hand, and, as part of this more general deficit, are sometimes overwhelmed by minor distractions. For example, a client who had been an audiologist complained that she could no longer meet a friend for lunch in the fountain court of a shopping center because she found the noise of the falling water to be an intolerable distraction. Prior to her accident she would have found this absurdly minor, but now it was overwhelming; with her specialized knowledge of hearing, this was a poignant difficulty. Other clients say they no longer enjoy social gatherings of more than two or three people because "I can't stand the noise." The real problem is reduced attentional selectivity, which turns multiple conversations into a cacaphony. Social isolation, which is one of the hallmarks of significant traumatic brain injuries, has one of its origins in this phenomenon.

## 2. Changes in Personality

The hallmarks of personality change after brain injury are irritability and bad temper, together with a cluster of socially inappropriate or personally dangerous behaviors. J. M. Harlow won himself a fixed place in the brain injury literature because of his pithy yet extraordinarily evocative description of the case of Phineas Gage,[4] who, on September 13, 1848, had a tamping iron blown through his prefrontal lobes:

> He is fitful, irreverent, . . . impatient of restraint or advice when it conflicts with his desires, at times pertinaciously obstinate yet capricious and vacillating, devising many plans for future operation which no sooner are arranged than they are abandoned in turn for others appearing more feasible. His mind was radically changed so that his friends and acquaintances said that he was no longer Gage. (Harlow, 1868, cited in Damassio, 1994, p. 8)

There are two inexactitudes in this classic description. The first, sadly, is that persons who have sustained significant personality change are still themselves—but with some personality traits grotesquely heightened and others partly obscured. Like someone you know who is wearing a mask,

[4]In Damassio's (1994) wonderful retelling, Gage is the centerpiece of modern-day detection, and "no longer Gage" because his emotionality had been disrupted.

there is enough of the old person left to make the change even more disturbing: Gage is still Gage. The second difficulty is that the lucidity of Harlow's description suggests that there is an entity called the frontal syndrome, a term that has had more vogue than it deserves. There might be a frontal concatenation, but not a syndrome: The unpredictability of the combinations in which frontal dysfunction appears is its most invariable feature. For the purposes of behavioral description, the only useful course is to describe the discrete manifestations of hypofrontality that are unique to that particular case.

## Anger

The open and appropriate expression of aggression is a prerequisite for adaptive behavior in both animals (Lorenz, 1963) and humans. Unfortunately, the adaptive use of aggression readily becomes maladaptive (Nell, 1990b, 1990c). Interpersonally, expressive anger is an immediately available means of establishing dominance and control, expecially useful to a frightened or insecure person in reestablishing some predictability in a seemingly chaotic world. Such behavior can rapidly become intolerable: It is corrosive of relationships, destroying family and social life, and is thus always a priority for therapeutic intervention.

In order to assess the threat dyscontrol poses to family and work relationships, it is useful to grade dyscontrol on three dimensions: *intensity*, *frequency*, and *spread* (Table 7.1). High intensity and freqency invariably result in institutionalization; high freqency alone, even with moderate intensity and spread, is corrosive of family life, and, if chronic, will often result in separation or divorce. A wide spread, even with relatively low intensity, is compatible with acceptable social or vocational adjustment. A case history gives the flavor of moderately severe dyscontrol:

> This young woman was unconscious for six weeks after a motor vehicle accident at age 24 years, having been admitted to hospital with a Glascow Coma Scale of 7/15. She was seen initially two years posttrauma, and at that time complained of increased "rattiness" with her son, then two years old, and her boyfriend, with whom she was sharing an apartment. The EEG was normal. A year later, she was referred back by the treating neurosurgeon because of increased aggression and inability to cope at home. She was seen with her boyfriend. They said that temper outbursts, directed sometimes at him and more often at the child, were a daily occurrence. Most were confined to verbal abuse, but at irregular intervals, usually weeks apart, she would throw food dishes to the floor, smash objects, and storm through the apartment, slamming doors. Two months earlier, she had "snapped" when the child answered her back, seized a belt and beaten him violently. She states that she "went blank," and her recall for this episode was patchy. This account is consistent with other accounts of an "altered state of consciousness" during

TABLE 7.1
Grading the Severity of Aggressive Outbursts

| *Intensity* | *Frequency* | *Spread* |
|---|---|---|
| | | *Confined to the home and to:* |
| 1 Shouts | 1 Two or three times a year | 1 Most significant other only (spouse, parent) |
| 2 Uses bad language | 2 Once every two or three months | 2 Other family members, pets |
| 3 Bangs objects around, slams doors | 3 Once or twice a month | 3 Non-family in the home (maid, visitors) |
| | | *Outbursts in other settings* |
| 4 Throws objects at wall or floor | 4 Once or twice a week | 4 Homes of other family or friends |
| 5 Smashes objects and furniture | 5 Three or four times a week | 5 Institution (school, hospital, residential institution) |
| 6 Threatening violence to people | 6 Once or twice a day | 6 Workplace |
| 7 Throws objects at people | 7 Many times a day | 7 Public places (street, bars, clubs) |
| 8 Attacks with hands or light objects: injuries none or slight | | |
| 9 Attacks with fists: significant injury, no medical care needed | | |
| 10 Attacks with fists or weapon: significant injury requiring medical care | | |
| 11 Life-threatening attacks with fists or dangerous weapon | | |

dyscontrol episodes—the subjective sense of being out of control, the patchy amnesia, and the interruption of "life flow." She was then tried on carbamazepine to which she at first responded well, reporting diminished rage frequency and intensity, but increased drowsiness. At the same time, she entered a group therapy treatment programme for brain injured persons. Some eight weeks after starting medication, another major outburst occurred, this time on the way home from a group session. Her son, in the back seat of the car, angered her and she turned around and beat him with her fists until the child was pulled away from her by her mother. Again, she was partially amnesic for this episode. She then terminated treatment.

Underlying the dyscontrol syndrome and contributing to it is a pervasive sense of confusion and uncertainty that arises from the fragmentation of the individual's inner world. This existential pain is only occasionally captured in the vast literature on brain injury. For example, Bach-y-Rita, Lion, Climent, and Ervin (1971) noted that for dyscontrolled individuals, there is frequently a sense of impotence in dealing with the environment, giving rise to a fear of impending total loss of control, freqently verging on panic. They maintained that small variations in the environment provoke massive repercussions, echoing Goldstein's (1942) observations on the catastrophic reaction seen in the brain-injured soldiers with whom he worked, elicited, for example, if an orderly rearranged a patient's items of clothing in his cupboard.

In conclusion, it must be noted that dyscontrol is a neglected syndrome. Although large case samples have readily been assembled by workers in this area (700 cases by Monroe, 1975; 286 by Elliot, 1982; 130 by Bach-y-Rita et al., 1971; 138 by Mungas, 1983), suggesting that its incidence rate places it among the larger mental health problems, its epidemiology remains unknown (Monopolis & Lion, 1983; Woody, 1988). By this neglect, psychologists and especially clinical neuropsychologists are failing to meet the diagnostic and treatment needs of a profoundly troubled client population, and dissipating research opportunities that hold the promise of shedding new light on a profound human problem, the pleasurable release brought by violent aggression.

## Tact and Social Appropriateness

Although hypofrontality has significant cognitive outcomes (see later), it takes its major toll in social relationships. Parents, spouse, children, and workmates of the frontally depleted person will complain sooner and with more feeling about aggression, tactlessness, and socially embarrassing behavior than about impaired memory and muddled problem solving.

### *Speech Pragmatics*

Speech pragmatics, defined as the study of language in context, and especially the speaker's ability to adapt to the demands of the current context, are especially vulnerable to frontal depletion. Penn (1988) quoted the saying that closed head injury patients "talk better than they communicate" and noted that standard measures of communicative competence fail to identify failures of speech pragmatics such as topic maintenance, topic introduction, and turn-taking behaviors (e.g., pauses, interruptions, and feedback to the previous speaker).

Mr. L, formerly a loss adjustor in the aviation industry, was unconscious for 10 days after a motor vehicle accident, with a 3-month period of posttraumatic amnesia. Here, 15 months postaccident, he is responding to a question about the normal duration of a conversational turn:

> The single part is probably, again it depends. I am an expressive type of person, so therefore I would say that my turn has to be longer than anybody else's. I'm now free to run away with time and words as they come into my head. I want to show people that I have the ability to be able to create correctly structured sentences, and that they will sit back and say, "That is correct." Yes or no, whether they will agree with it or not is immaterial. The

> point being is that they have heard it and they have had to think about what has been said, as opposed to me burbling nonsense which they can turn round and say, "That's a load of garbage."
>
> I won't say again, you know again the solidity of what I want to say is vaguely walking away from it. . . . When I feel as I do at this moment, that I must make a point, and be solid about the point, that time becomes immaterial to me, and that I might capture you in a corner and go at you for two or three minutes, and I won't look at my watch until I'm sure that you . . . are aware of the point that I'm making. [Later in the conversation Mr. L comes back to this point and says:] I can go on forever until I'm certain that the person has derived knowledge of the point I'm putting across.

This remarkable passage captures Mr. L's fear that his listeners will turn their backs on him because they think he is talking nonsense. His press to speak—"to capture you in a corner and go at you for two or three minutes"—is to make quite sure that he has been understood, that he has been "solid about the point." He is loquacious to show that he can create well-structured sentences that will make people sit back and say "that is correct" whether or not they agree with it.

The *hyperverbality* often seen after traumatic brain injuries arises in this fear, and the belief that if a point is made at length and repetitively, people will at last not only understand but also respect the speaker. The fear that this may not be achieved is acutely present. At the end of the session Mr. L says reflectively, "I tend to be suffering from long-winded garbage." The comment typically heard from caregivers, more aware of their own impatience than the speaker's striving for respect, might be, "He used to be a good listener, now he just overtalks."

*Sponginess* further increased the wordiness of Mr. L's speech flow. Each successive formulation just misses the point. For example, wanting to say that he will not time his own utterances, Mr. L begins as follows: "I won't say again, you know again the solidity of what I want to say is vaguely walking away from it. The single part is probably, again it depends . . ." Crisply, he would have said, "I made this point before." Both hyperverbality and sponginess are attempts to conceal cognitive fragmentation.

*Tangentiality* is a cognitive problem that originates in the loosening of frontal inhibitory control, ranging from some discursiveness to a florid billiard-ball conversational style, bouncing associatively from topic to topic through a train of clang associations and semantic red herrings. Despite his verbosity and muddle, Mr. L stays tenaciously on the point, but when tangentiality in its more extreme forms is added to this mix, speech flow becomes near impossible to follow:

> Asked for his last recall before the accident, a client who had sustained severe frontal injuries while on a camping holiday said, "We were in the

pickup on the way back from lunch and there was this huge snake on the road, it crossed the road just in front of us, it's very dangerous, Jaap Cilliers, this man I work with, his brother died from snakebite, Dr Albert gave him injections but it was too late, he's our doctor too, this pain I was telling you about, he told my wife that if he didn't operate I would die in a month . . ."

***Discourse Register.*** A myriad social rules, considered in more detail under "Tact" later, determine the discourse register appropriate to people and occasions. Some languages distinguish between the formal *Sie*, *vous*, or *U*, and the intimate *du*, *tu*, and *jy*: What one says to a "du" is by no means welcomed by a "Sie." One uses first names to the one, but not without permission to the other, swears in the presence of one's mates but not one's boss or clients, expresses one's irritation to the former but rigidly controls it in the presence of the latter.

A 28-year-old woman who had a post-traumatic amnesia (PTA) of 3 months following a motor vehicle accident was employed 5 years postaccident as the organizer of a new product launch. Her employer was interviewed with the client's permission:

> I found M incredibly difficult to work with. She had a huge ego. She was extremely confident, she addressed people in a familiar way, she used a lot of four letter words whether she knew people or not, these were eminent people, big names, and they didn't like it. She made a call to one of our VIPs and forgot to ask for his address, so she immediately picked up the phone again. I said to her, "I've got his address on file," but she went ahead and phoned anyway. He was really irritated, he's someone you don't treat lightly and M knew it. He said to me later he was sick to death of our company and the endless telephone calls.

The huge ego and arrogant confidence are masks for M's fragility and pervasive sense that things are slipping out of control. For example, another complaint by the employer was that she had forgotten to pick up a VIP and bring him to the launch as arranged; he had arrived by taxi, late and irritated. Asked about this at the interview, M said, "Oh no, I didn't forget. I wrote everything down. I had a briefcase full of things I had to remember, I was right on top of things all the time." Again, personality and cognitive changes are interwoven, but governed by the same ego-protective mechanisms.

***Social Norms.*** A problem related to discourse register is responsivity to the nuances of social context. Social occasions have culturally determined parameters for dress, solemnity, and discourse content. The publicity agent already described further angered her clients by appearing at the launch party, as the employer put it, "in a sweater with threads hanging out of it." These dress and conduct parameters are quite rigid for ritual occasions such as christenings, weddings, or funerals. At a workshop series on the diagnosis and treatment of brain injury that our unit gave for primary health care nurses in the Baragwanath Hospital community clinic system in Soweto, Johannesburg, one of the nurses listened attentively to our review of impaired social norms, and then told the following story:

> Her nephew had at the age of 18 driven his car into a truck. His sister, a newly qualified nursing sister, was killed outright, and he had been unconscious for five weeks. Since then, he had several times returned to his profession as a teacher, but had been dismissed on each occasion. He was careless of his personal hygiene and appearance, wearing dirty clothes and trousers with holes that showed his underwear. At his mother's funeral three years previously, he had shocked the gathering by telling jokes at which he laughed uproariously.

#### *Personal Hygiene*

As the previous case history suggests, personal hygiene sometimes falls victim to frontal damage. On the one hand, such individuals are no longer concerned that their appearance or body odor may give offense. But they are also so distractible and stimulus-bound that they cannot formulate or carry out the intention to bathe, and even if they do, fail to wash properly. A mother will say, "You have to stand there and watch him, otherwise he just splashes the water around and gets out." The wife of the afrontal instrument maker described in the next section complained that although he was formerly fastidious, he would not change his underwear unless reminded, and often claimed to have bathed; however, the bath towels would be dry and unused.

### Affect and the Perception of Emotions

Failures of empathy and sensitivity are more common after right hemisphere damage, but in the aftermath of acceleration–deceleration injuries, these lapses are inseparable from the other behavioral markers of frontotemporal damage. Thus, insensitivity to the feelings of others is a consequence of a cognitive inability to experience one's own feelings fully and deeply. A close friend dies and the response is "Oh shame," danger brings no sense of fear, and events that customarily give rise to intense joy or sorrow leave the person unmoved. For example, consider the following account:

> A business executive who had become adynamic and unable to make even the simplest decisions said, "Tracy and I decided to sell up and move down to the coast. That's something we'd planned on doing for years. Tracy was over the moon, but it left me cold. I bought a skiboat and we went out to sea,

> it was like a dream coming true, but I couldn't feel a thing. Our granddaughter is coming to stay with us, I know I should be happy but I'm not really.

Sometimes the appropriate emotion is verbalized but remains unfelt:

> A 48-year-old journeyman instrument maker, a trade requiring a high degree of precision, was having a social drink with his wife in a Johannesburg bar when an allegedly deranged karate expert struck him full in the face. He was seated on a high stool at the bar counter, and fell backwards from this full height, striking the back of his head violently against the floor. He was comatose for 5 weeks, with fluent aphasia that thereafter appeared to resolve. A major behavioural disturbance with rage dyscontrol then manifested, as if his frontal lobes had been surgically excised.[5] His wife reported that "when he gets angry, he goes out of control. Last week he grabbed me by the hair and started banging my head against the floor. When I screamed he let go, then he began weeping and saying how sorry he was and begging forgiveness, then a minute later he'd forgotten all about it. He says he's sorry but he isn't really."

As also in this case, formerly loving sexual partners become peremptory and oblivious to their lover's feelings or needs. Family relationships are affected. A young woman who had a close and caring relationship with her older sister became coarsely critical of this woman and her husband after a motor vehicle accident. When the husband was retrenched and unable to find another job, she would say to her sister in this man's presence, "You should never have married that useless loafer. Just look how he sits around all day sponging off you."

***Fear and Risk-Taking.*** Like joy and sorrow, fear may also be attenuated. After his recovery, a young motorcycle mechanic resumed his custom of riding his own bike several miles to and from work along a busy highway: "I go too fast, I jump traffic lights, I overtake on the inside, I take bad risks, but I'm not afraid. Before if I rode like that I was scared, it was a thrill, now I feel nothing." Diminished fear of injury is a predisposer to further injury, explaining the epidemiological finding that the incidence rate for a second head injury is three times higher than for the first, and eight times higher for the third (Annegers, Grabow, Kurland, & Laws, 1980).

## Tact

***Confidentiality.*** Information in the public domain—what the newspapers say, company results—is freely exchanged. But there are precise culturally mediated rules about the disclosure of private information, such as one's

[5]This is consistent with a heavy fall onto the back of the head, which sets up a low pressure cavitation effect at the frontal poles (Pang, 1985).

own financial status (disclosed to financial professionals, equal-rank business colleagues, or close adult family), or relationship problems (best friend, mental health professional). Brain injury may undermine these confidentiality markers:

> A young man who had been comatose for 3 months after a motor vehicle accident made an excellent physical and intellectual recovery but continued at 2 years postaccident to cause himself and others embarrassment by socially inappropriate behaviour. At a family dinner party, when the conversation turned to a movie about the breakdown of a marriage, he told the amazed guests, "My parents have separate bedrooms. They haven't slept together this whole year." Later in the evening, he disclosed the exact (and very substantial) profit his father had made on a property transaction that morning.

***Timing and the Development of Friendship.*** The culture prescribes clear markers that define the pace and timing at which a friendship develops, the changing level of intimacy acceptable at each developmental stage, and the frequency of contact between friends. These norms may be disrupted by brain injuries with a frontal component, overwhelming new friends with unwanted intimacies. At the courting age, at which most brain injuries occur, deviation from social norms is not tolerated:

> A young man with a profound frontal injury spent his days in shopping malls casting about for a likely mate. "I met this bird outside Woolworths and we get talking and I say to her let's have some coffee, so we're having coffee and I ask her if we can go steady, but she says no, she has a boyfriend, then she says she's late and she goes off. She didn't even finish her coffee."

Both men and women whose self-image has been subverted by physical impairments or scarring find the prospect of an intimate relationship deeply comforting, and may mistake exploitiveness for commitment; women disinhibited by brain injury are at especially high risk of sexual exploitation and unwanted pregnancy.

***Social Perseveration.*** The following vignettes give the flavor of this problem:

> A 16-year-old unconscious for 5 days after coming off a motorised bicycle and striking his head against the kerb tickles his sisters and pulls their fingers when they come to visit him until they get angry.

> A strapping 15-year-old, proud of his strength, would show his affection to visitors by pinching their arms, continuing with this unwanted attention until they left, refusing to return.

> At a New Year's party, a young man, whenever addressed, would throw his hands up in mock consternation and say, "Don't blame me. I'm not guilty." Despite an initial warm welcome, he found himself isolated and ignored after he had behaved in this way for several hours.

The underlying confusion and distress are clearly visible through these screens of pseudo-euphoria.

*Loneliness* and *social isolation* are common consequences of TBI. During hospitalization and early convalescence, friends shower the victim with attention, but as changed physical capacities and personality changes become apparent, ruling out previously shared sporting and social activities, friendships falter despite the best of intentions. Close friends may maintain contact, but visits become more widely spaced and may cease if there is aggression, hyperverbality, adynamia, or other major personality change. Adynamia is a further contributor to loneliness:

> An avid squash player who had made an excellent physical recovery after a 3-month coma was puzzled that his friends had stopped calling him to arrange games, but never himself initiated such arrangements.

> A nurse at our Primary Health Care workshop remarked of a colleague injured in a heavy fall, "She used to be the bright spark socially, now she just waits for someone else to do something."

***Suspiciousness.*** Misreading social contexts and misreading the emotions of others are closely related. A harmless question, "Did you bring back the change from the shop?" will unleash a vituperative rejoinder, "Are you calling me a thief?" One is tempted to describe this inability to read affect as "emotional aphasia," which may in turn lead to a chronic vigilant suspiciousness. In the following case, the young man's confusion is exacerbated by the misperception of social cues; his response to the confusion is to find release in episodes of violent dyscontrol:

> A 23-year-old man had been unconscious for 3½ months after a motor vehicle accident 15 months previously. Presenting complaints by the mother and stepfather, to which the client listened with approval, were of violent outbursts several times a day, during which he would bang his head against cupboards, smash his fists into furniture, and swear at his mother, sister or elder brother—but not his stepfather. As rage gripped him, his colour would heighten, and the verbal abuse would be accompanied by a threating physical display, for example, advancing on his mother or brother while shouting, "Hit me, hit me, put me in an asylum," or on one occasion banging his head against the kitchen wall and violently biting a tap handle. One of these outbursts was in error recorded on my answering machine, and the dysphoria—even panic—was very striking. However, in listening to his

mother's account of an outburst at a weekly psychotherapy session, he would smile, nod, and look gratified rather than remorseful. Clearly, the dyscontrol outbursts generated a significant sense of inner release.

Whatever neuroendocrine mechanisms might have been contributing to this gratification through rage, it also became clear that the outbursts were an attempt to reassert autonomy and control in a world he could not understand. For example, when asked, "Why did you shout at your mother this morning?" he was unable to recall any precipitating factor, saying instead, "They musn't take chances with me." This phrase was a blanket response given to all infringements of his dignity, and was invariably advanced as an explanation for dyscontrol episodes.

A Thorn in the Flesh

No single case example can capture the complexity of the interwoven cognitive and personality changes that follow a severe closed head injury—but some come close:

> Selby Moshidi, 34, a former teacher, personnel officer and successful life insurance salesman, had been struck by a bus travelling at high speed while on an early morning jog in Soweto, Johannesburg. He arrived at the assessment session—and for all subsequent meetings of the therapy group we invited him to join—in a colourful velvet and print smoking jacket, and having listened with interest to his mother's description of her problems with him, said, "I am a thorn in the flesh of my mother. Now I must go to the toilet to answer a call of my digestive system." His mother, 59, complained that he was hyperverbal, changing subjects unpredictably (in her words, "he jumps topics"—though at other times, she continues, he repeats one sentence all day), inconsiderate, getting up in the middle of the night in a house with 12 other people to cook food. On other occasions, he goes out at night to shout at the neighbours that their radio is too loud; the neighbours then emerge and assault him. He and his mother were accompanied to the interview by their next door neighbour, who says that her kitchen looks into theirs. She says that Mr Moshidi used to be competent at home but now leaves food to burn and scorches the clothes he attempts to iron. The mother says that he had been a kind man who loved his child, but now she cannot leave the house without being worried about both him and the child. She says he becomes childish when he is angry, "going on hunger strike for a week." He speaks only English, eschewing the African vernaculars used by everyone else in Soweto. She says that his friends have abandoned him, wearied by his verbosity and unpredictable moods. Sometimes he is happy and laughing, making jokes constantly, and at other times ill tempered and vindictive. When his nephew visited and used the bathroom, Mr Moshidi became first impatient and then enraged, finally battering at the door and attempting to stab the young man with a garden fork.

### Emotion as a Motivational Driver

It is not by chance that emotion and motivation share a common root in the Latin *emovere*, to move out: both mobilize and drive. The reason for beginning and persevering with a difficult job—overhauling a car engine, writing a neuropsychology book—is the sense of joyous anticipation at the outcome, a premonitory thrill of achievement. Guilt (which, like joy, is another great driver) has the capacity to sustain goal-directed activity even when it has become tedious to do so. Recall the business executive who retired to the coast: His adynamia and his emotional flattening, to which neither going out to sea in his own boat or his granddaughter's arrival bring joy, are not two separate problems: The flattening *is* the adynamia.

## 3. Cognitive Changes

Whereas the hallmark of personality change after brain injury is anger, ranging from irritability to rage, the hallmark of cognitive change is unreliable memory. These are not present by happenstance: The first is an expression of disinhibition and the second of frontal distractibility (which may be amplified by axial memory disturbances). Both disinhibition and distractibility are in turn expressions of the same underlying problem, a reduced level of arousal. Both anger and memory difficulties, because of their origin in a more general underlying disruption of cerebral functioning, are linked with many related changes in personality and cognition (sometimes as the cause of these changes, and sometimes as symptoms).

### Memory and Arousal

Many of the puzzling aspects of memory after a period of unconsciousness are explained by one simple phenomenon: Memory is vulnerable to reduced arousal. One does not need elaborate Freudian hypotheses about ego defenses to account for the evanescence of dreams: At the point of transition from sleep to waking, arousal is low and memory poor. Dreams fade unless one takes extraordinary steps to preserve them, such as a notebook at one's bedside and a fixed determination to record the fading images even before one's eyes are fully open. Similarly, the last page of the book laid aside as one's eyes close in sleep is lost from memory by morning, and so are the news headlines playing on the clock radio alarm while one is still half asleep—though one has tried hard to attend fully. The links between memory and arousal are vividly illustrated by clients who cannot remember yesterday's dinner party but recall in vivid detail what happened when, for example, the house caught fire last year. Other distant but dramatic, and therefore arousing, events may also be remembered:

> A 42-year-old man would forget information that had been exchanged a few minutes before in the therapy session, but could tell—and retell—the smallest details of a sailing incident five years before, long after he had sustained his brain injury, when he and his companion had nearly drowned when a storm came up.

Families are puzzled at these apparent inconsistencies, and end up saying in frustration, "He only remembers what he wants to remember." It would be more accurate to say, "He only remembers what excites him."

### Posttraumatic Amnesia: The Puzzle of Variable Duration

The otherwise inexplicable course followed by returning memory after a concussive injury is also largely explained by the mechanisms of arousal. The first puzzle is that although consciousness returns, memory does not (as with the awakening sleeper). The second puzzle is that although continuous memory may appear to have returned, this may not be the case. As a result, the duration of PTA typically increases with passing time (discussed later). In the emergency room, a TBI victim may say, "I woke up lying in the road with people around me"; next day on the ward, the same person in all honesty reports a first recall of waking up in X-ray. A month later, first recall may be for an event that took place a week or two posttrauma. One period of PTA thus becomes nested in another. This is not "dishonest." Six or 12 months posttrauma, by careful questioning of the client and caregivers, it is possible to establish the point at which continuous memories begin to be laid down; this point tends to be consistent across examiners and can often be verified by other witnesses who are in regular contact with the client.

Another difficulty in determining the duration of PTA is the problem of "island memories." Clients will often relate an episode of vivid recall embedded in a longer blank period, for example: "I can't remember the impact, then I remember waking up under very bright lights and a man in green standing over me. Then I go blank again until I woke up in the ward. The nurses told me I was in an accident three days before. From then on I can remember everything as it happened." When does PTA end? Does it end at the time of the island memory, presumably in the emergency department or soon thereafter? Or does it end 3 days later in the ward? By definition, PTA continues until the time from which consecutive memories are laid down. From that point, the patient can for example remember what happened the previous day or the previous week—so an "island memory" does not mark the end of PTA (Nell, 1997b).

With the severe memory disturbances sometimes encountered after brain injury, PTA imperceptibly merges with the chronic memory disorder so that it is difficult to say when the one stopped and the other began. However, as Wilson, Baddeley, Shiel, and Patton (1992) have shown, the core mechanism of PTA is not only amnesia, but also hypoarousal. The latter gives rise to slowed comprehension, increased comprehension errors, slowed word generation, and slower reaction time. Specialized testing can therefore reliably distinguish between PTA and the chronic memory impairment that may arise from brain injuries.

### *The Reliability of Posttraumatic Amnesia as a Measure of Severity*

Some clinicians who pride themselves on their tough-mindedness are contemptuous of PTA because of its early variability, and dismiss it as a "useless measure." This view is eagerly embraced by the defendants in personal injury matters. They will seize on the fact that the plaintiff was awake and talking coherently 3 days after the accident as hard proof that he did not have a PTA of 3 months, as claimed by the plaintiff's neuropsychologist. Such views reflect expedience or ignorance. In fact, as previously noted, once the final duration of PTA is established some months postaccident, it remains strikingly stable. Its early variability reflects not its "unreliability," but the client's struggle to reestablish continuous memory in the face of reduced arousal. It is for this reason that PTA remains one of the best indicators of the severity of a traumatic brain injury.

### *Arousal-Dependent Memory Deficits*

Selective remembering is the hallmark of these difficulties, well-captured by the client discussed earlier who could remember only what excited him. The caution that arises is that testroom memory performances cannot be extrapolated to the schoolroom and workplace without making full allowance for the fact that arousal levels in these routinized and therefore boring settings are typically lower than during testing.

A related problem is shallow encoding and subsequent difficulty in retrieval. Clients often use the metaphor of a filing system, saying that they can no longer keep the boxes in their heads tidy, and cannot remember in which box they have stored a particular piece of information. Copious note-making may then be tried, but may not be successful because of the same muddle. For example, a logging company executive complained that he kept detailed lists of things he had to do, but they became so complicated (he described these lists with their maze of arrows and colors as a rat's nest) that he could not use them.

Retrieval from long-term memory may be affected. Mr. L used the metaphor of card files:

> You write down the information about a particular client or problem and then you will store it away for the years that you need to and then you say, "Right!" and you can have a look and see the particular details about that person. The problem I'm having at the moment is . . . that there are thousands of card file systems that I haven't got access to open them and pull out one or the whole stack of them under that name or initial. I see boxes of card systems inside my head which tell me that the information is there but my brain is preventing me from having access to specific information about any of those boxes.

*Quasi-mnestic* difficulties arise from distractibility of frontal origin. This can give rise to problems in staying on task long enough to hold on to the necessary information. Unlike axial amnesia (discussed later), these are not primary memory deficits, but secondary to hypofrontality. The next example gives the feel of this "forgetting to remember":

> Bone sepsis after a ruptured aneurysm of the anterior communicating artery had virtually ablated the prefrontal lobes of J, a 16-year old schoolboy with a verbal IQ of 120 (PIQ 85, FSIQ 102) on an age-appropriate South African scale. He was asked what half of one-third of the height of the Eiffel Tower would be if the tower was 150 meters high.

*J.* Aha. Half of one-third, give, give me, one-third of the actual height. And, and then you—I don't know the answer you get for that one.

*Examiner.* How high is the Tower.

*J.* I haven't got a clue.

*Examiner.* It's 150 meters.

*J.* Oh yes. 150 meters. (Silence)

*Examiner.* What's half of one third of the height? . . .

*J.* Half of 150. Two point, 2½.

*Examiner.* How do you get that?

*J.* Because half of, half of 5, will give me 2½, is equal to . . .

*Examiner.* I'm confused. Where does the 5 come from? Do you remember the question?

*J.* The question was that the Eiffel Tower is, the Eiffel Tower is, is, um . . . The Eiffel Tower is . . . Scheeze!

### *Axial Memory Deficits*

These primary memory deficits are quite unlike the quasi-mnestic difficulties considered thus far. They arise from damage to the limbic structures within the temporal sulcus and result in partial or complete inability to tranfer new information from echoic memory to long-term storage. In their extreme form, such deficits are seldom seen after traumatic brain injuries, but may manifest as a partial amnesic deficit that is additive to the arousal-linked problems described in the previous section.

Just as arousal-dependent memory troubles caregivers because of its apparently wilful quality, problems also arise within the family because of the relative hardiness of remote memories. A valuable therapeutic contribution to family functioning is to explain that the memory problems seen after a brain injury are for new learning and arise from failures of the mamillo-limbic "print now" mechanisms; old information laid down and consolidated pre-injury is typically spared, such as the details of previous employment or residence, complete with the telephone numbers of long-vacated homes.

***Procedural Memory.*** A common error—especially in adversarial assessments—is to infer that if test–retest improvements are found after spontaneous recovery has ceased, learning and long-term retention are retained, and the memory deficit described on interview or revealed by testing is the result of role enactment or worse. There is a large literature on procedural as distinct from declarative memory, to which such improvements must in part be attributed: Just as postencephalitic amnesics can by repeated practice acquire great proficiency at a task they cannot recall ever having seen before, so may amnesic clients improve their Block Design, Object Assembly, or Digit Symbol scores with little if any declarative recall for previous administrations.

## Problem Solving

Vigorous and reliable memory is fundamental to every aspect of everyday functioning and also to all problem solving.

***Working Memory.*** Working memory is short-term memory (STM) in action (Baddeley, 1986). It is active in the service of problem solving by processing material drawn from STM, and returning it to storage after manipulation. Its likely component processes are speed, capacity, and rate of forgetting (Necka, 1992). The most-used test of working memory is Digits Backward: The forward series must be held in STM while it is reversed, in precise agreement with the definition given. All multistep problems draw on the same capacity—the answer to the first part of the problem is held in STM while it is further manipulated to arrive at the final solution. A common error is to assume that such problems are restricted to the arithmetic modality: In fact, all verbal or figurative analogies problems, as well as complex tasks such as Raven's Progressive Matrices, the mental rotation of three-dimensional figures, and the interpretation of mechanical or architectural blueprints, draw on these capacities to store, manipulate, retrieve, and restore information.

Persons with working memory difficulties will typically make nonspecific complaints—that they take longer to work things out, cannot do mental arithmetic, make more mistakes than before, or solve problems in a roundabout and needlessly time-consuming way: These difficulties can arise from primary memory failure, poor concentration, distractibility, and any number of other causes. Here, testing has primacy, because a differentiated description of the origins of these problems will require a test-based evaluation of the separate contributions made by these and other cognitive processes.

***Flexible Set Changing.*** Smoothly moving between differing performance expectations—from one cognitive set to another—is a prerequisite for effective problem solving. Without this ability, the person remains obstinately locked into an ineffective approach, trying to loosen a frozen nut with a pair of pliers, ignoring the penetrating oil and socket wrench that lie to hand.

Neuropsychologists have long been aware of the vulnerability of fluent set changing to head injury. Trail Making is one of the oldest tests in neuropsychology, having made its way from the Army Alpha to the Halstead–Reitan battery and thence to a place in most neuropsychological assessments. Like the Stroop, it requires suppression of an overlearned response in favor of a novel one. Other testroom analogues of real-life rigidity are difficulty in moving from color to form on the WCST, from number to position on Subtest 3 of the Halstead Category Test, or to alternative strategies on construction tests. The schoolboy at his homework repetitively applying the wrong theorem to a geometry problem is found on a formboard test to push and bang at an assembly that has been incorrectly begun, hoping by force to make the remaining pieces fit into the space that is left: The necessary step of trying a different approach—at school or during assessment—is not taken.

***Faulty Analysis.*** In the workplace, the most ubiquitous form of problem analysis is "troubleshooting." At the simplest level, there may be difficulties in using appliances (i.e., struggling to replace radio batteries correctly, or flipping an audiocassette on its long axis rather than the short axis when changing tape sides). Again, some case anecdotes will clarify the the problem:

> A 48-year-old photocopy machine repairman found that he could no longer trace machine faults efficiently; part of the problem was that he could not interpret wiring diagrams correctly, though he had been skilled at this previously. The employer, taking account of his long service, moved him from repair work to training apprentices. He said with some bitterness that "these kids" would often find the answer to a problem before he did. The same difficulties arose at home. He said that when the front door bell stopped working, he had needlessly stripped the wiring before checking that the transformer was plugged into an outlet.
>
> The young motorcycle mechanic who had lost his sense of danger [see "Fear and Risk-Taking" earlier] had been the repair shop's electrical expert before his accident, but was now anxious and frustrated. He said on interview that the previous day he had reassembled the wiring harness of the bike he was repairing. With only a few connections remaining to be made, he found that the headlamp would not illuminate. Instead of tracing this fault to its source, he had ripped out everything he had done to that point and started again.
>
> During the test session that followed, having done very well on Block Design, he was working on the last and most difficult item. With seven blocks in place, he struggled to orient the last two blocks needed to complete the design, then, overwhelmed by task overload—a phenomenon Kurt Goldstein called the catastrophic reaction—he swept the entire array away and started again from scratch, replicating his on-the-job method of coping with overload.

Faulty analysis errors that emerge in stories like these are detected on testing by qualitative analysis of performance on mazes, the Complex Figure Test, Raven's Progressive Matrices, the Wisconsin Card Sort, and constructional tests.

### Motivational Deficits

***Distractibility.*** The ability to stay with a task until it is completed may be affected. In its extreme form, this is a headlong flightiness like continued TV channel-hopping, abandoning one thing for another as soon as it is begun. More typically, there are complaints about getting distracted and then forgetting about the job at hand, for example leaving paint and brushes to dry in the sun because the dog came by and the painter went off to have a romp.

***Self-Monitoring.*** Task performance standards are maintained by feedback loops that continually check output quality; unacceptable segments are then revised to a standard acceptable to the individual. Without effective self-monitoring, task performance becomes unacceptable. Fellow workers say that the person in question was a perfectionist before, but now her desk is a mess and she does not check her typing for errors. At home, the complaint will be that whatever the person does has to be checked and redone (e.g., if they wash the dishes, they are spotted and greasy; or if they sweep the floor, streaks of grime remain).

Poor self-monitoring shows up on many tests; a specific method I have found useful is to administer the Geometric Design Reproduction Test (Fig. 10.5), and, if it is perceptibly sloppy or untidy, to say, "Are there any of these drawings that you're not satisfied with?" If the client indicates one or another of the designs, I ask, "Would you like to do anything about it?" Sometimes monitoring is present but the motivation to do something about it is absent. The formerly meticulous instrument maker already described said, "Sorry about that one" when he produced a triangle with a large gap at the bottom corner. Looking at his three attempts at the double-headed arrow, he said, "That's not a 1000%," and added, "But I'm not trying 1000%. I'm quite blase about it." Asked if he'd like to redo the design "to make it 1001%," he declined.

***Saying and Doing.*** A disjunction between saying and doing—faintly visible in the preceding case description—can mislead even experienced examiners into giving clients credit for maintained drive and self-initiated activity when closer examination would show this to be incorrect. Asked how he passes the time, a client may say, "I start off every morning with my exercise, at least half an hour, but I try to make it an hour, then I do my computer practice until lunchtime." If questioned, caregivers will say that most mornings, the client does no exercises, and if he does, for not more than a few minutes; during the computer practice period, they most often find him sitting on his bed staring into space. Consider another example:

A 22-year-old man who had sustained a brain injury during his army service with a reconnaissance unit was admitted to an inpatient rehabilitation unit. Asked about future plans, he said he was planning a rafting safari down the Orange River in north-western South Africa. He was physically strong enough to undertake this trip, and the occupational therapist provided him with topographical maps, requesting that he plan his day by day route. She offered to help him compile a list of equipment and supplies once this had been done. However, he made no use of the maps, though he continued to talk about his plans for the safari. He was encountered two years later at a support group for brain injured persons, where, asked about his future plans, he spoke with undiminished enthusiasm about rafting down the Orange River.

If the saying–doing disjunction is present, then all information that affects prognosis, treatment, or compensation should be checked with independent observers.

### 4. Physical Impairment

In well-developed settings, physical workups will be carried out by multidisciplinary teams in rehabilitation departments. Matters are different if the neuropsychologist is the first or only examiner, or in the overworked and understaffed conditions typically found in the developing countries. Here, if neuropsychologists with their special knowledge do not draw the attention of other disciplines to physical impairments, these may not be documented, depriving the client of compensation entitlements and whatever treatment may be available.

In these circumstances, it is essential that in the course of the neuropsychological examination *sensory function* should be examined with regard to olfaction, hearing, vision (especially the presence of field cuts which, unlike diplopia, may go unnoticed by the client), and peripheral sensation (e.g., clumsiness and motor slowing on the Grooved Pegboard may be secondary to sensory loss in the fingertips). Regarding *motor function*, the client's gait and balance must be observed, resting or intentional tremor elicited, a handwriting sample obtained, and fine motor dexterity examined bilaterally by a pegboard task and a paper-and-pencil targeting test (such as the Pursuit Aiming Test included in the WHO–NCTB). Physical problems are often overlooked during psychological examinations. But a psychologist who gives a clean bill of occupational health to a client with a brachial plexus lesion or a dense hemiplegia has been professionally neglectful, to say the least!

## PROGNOSIS UNDONE

The only clinical certainty about the course of recovery from a brain injury is its uncertainty. Most individuals who sustain mild brain injuries recover completely, but some are left with chronic headache and permanent intellectual debilitation. Most moderate and severe brain injuries leave indelible marks, but some individuals, even after prolonged coma, recover almost completely. Statistically, there are certainties, but a mean is one thing and an individual outcome is another.

Unfortunately, the course of *spontaneous recovery* after a severe brain injury creates some misleading certainties in clients and caregivers. The gradual restoration of memory is one aspect of the wider process of spontaneous recovery, which is very rapid in the days and weeks following the return of consciousness: Awareness of the environment is restored, drowsiness remits, daytime sleep decreases, and reality contact improves.[6] If cognitive functioning is serially tested in the first year postaccident, memory and problem solving show dramatic increases in the first few months, reaching a plateau sometimes as early as 3 to 6 months, but more usually at about 12 months. Thereafter, gains may continue to be made, but they are minor compared to the spectacular initial improvement. This rocketlike trajectory—a steep liftoff followed by a near-invariant orbit—creates significant therapeutic problems, because both client and caregivers come to believe that the initial pace of recovery could have been maintained if only they had worked hard enough or had the right advice. Consider the following case:

> A client with severe spasticity and poor balance as the result of an accident five years before ceaselessly drove himself to exercise in the fixed determination that if he worked hard enough, he would be able to play football again and to fulfil another ambition, learning to skydive. The efforts by well-meaning neurologists and physiotherapists to explain the after five years, no further improvement in his physical condition was possible, fell on deaf ears. His response was that immediately after the accident, "The doctors told my mother that I would never talk again or walk again, and look at me now." His motto was "No pain, no gain," and he persevered in a punishing exercise routine. In the area where he lived, his windmilling gait as he ran, flailing his arms to keep balance, was a familiar sight.

## For Better or for Worse

As distinct from the early stability of cognitive function (although there are some startling exceptions), the skills that contribute to life adjustment

---

[6]Hallucinatory ideation is unusual, but can occur. A severely hypoaroused client in Madison, Wisconsin, who repeatedly dropped off to sleep during the assessment, complained that his house was being moved from place to place every night while he slept. An Italian engineer working in South Africa had an appropriately international hallucination, maintained during the first 3 years posttrauma, when my follow-up ended: He said that he had been instructed by a Swedish woman to dive into a Swiss lake, where there were many gold coins and dead people under the water, and write their names onto a sheet of metal she had given him. A South African accountant injured in a motor rally accident believed he was part of a spy network that communicated through the sewage system, flushing messages down one toilet to be recovered in another, far distant—a fetching metaphor for the operation of South Africa's security services in the 1980s.

(impulse inhibition, social judgment, the empathy needed to sustain a relationship) continue to change through the individual's lifespan, sometimes for better, sometimes for worse. A brain injury can change a raging tyrant who had terrorized his wife and children for years into a mild-mannered and affectionate husband. Within a stable and supportive family, accident survivors, although profoundly changed, may moderate their expectations and settle into a placid, rewarding routine. Conversely, the case histories already given show that flight of ideas, hyperverbality, and rage can transform a loving relationship into a nightmare, as the survivor, soured by anger, becomes steadily more shrill and violent as the years pass, until in the end institutionalization may be the only viable option if the family is to survive.

### Neurosurgeon as Victim: A Footnote

Regarding cognitive recovery, most readers will be familiar with one of the most intriguing cases in the brain injury literature, "Neurosurgeon as Victim" (L. F. Marshall & Ruff, 1989). Larry Marshall, currently chairperson of the departments of Neurology and Neurosurgery at the University of California, San Diego, tells in this chapter and in L. F. Marshall and S. B. Marshall (1996) how he had slipped while walking, struck his head on a rock, was unconscious for less than 30 seconds, and then got up and continued the walk, politely rejecting offers from neuropsychologist friends to check him out. For the following 18 months, he suffered ongoing memory retrieval problems with client names and journal articles, and would daily lose his keys and glasses; he added that he was rattier than usual with residents. At the 1996 Advances in Acute Neurotrauma Conference in Philadelphia, where he made a dazzling presentation, he said that his symptoms had completely remitted some 2 years postincident, and that at this point, some 10 years later, his subjective sense was of complete well-being.

In an ironic twist to the story, L. F. Marshall and S. B. Marshall (1996) recounted that some 10 years after his accident, his wife fell heavily from horseback, was unconscious and flaccid for 4 minutes, and profoundly amnesic for the next 40 minutes:

> She could not remember an extended telephone call with a close friend nor attendance, four days before, at a Barbra Streisand concert. She could not name the president of the United States nor the governor of California. [At 40 minutes] there was a dramatic sudden surge in recall and bewilderment at her reported inability to recall events immediately folowing the injury. (p. 4)

For the following 6 weeks she was severely fatigued, with a reduction of work efficiency. The authors hypothesized that L. F. Marshall sustained bilateral damage to the fornices giving rise to the prolonged memory disturbance, whereas his wife, who had been helmeted, sustained an injury that was localized to the brainstem and caused chronic fatigue. They concluded that these subtle case-specific variations are obscured in studies of patient groups, and deserve attention in their own right.

### Outcome Assessment

A prognostic difficulty is now moving from the psychological arena to the high-powered world of acute care neurotrauma drug trials. The pharmaceutical companies that fund such trials, each at a cost that may run into millions of dollars, are exerting increasing pressure on the research community to develop assessment methods that sensitively reflect subtle, high-level cognitive and life quality differences between treated and untreated groups. It is unlikely in the foreseeable future that test-based assessment will meet this stringent need, and increasingly, assessment will have to turn to quantifiable behavioral assessments to which the behavioral interview method and a variety of supplementary life quality questionnaire items[7] will contribute.

## THE GLASGOW COMA SCALE AS A PROGNOSTIC INSTRUMENT

The Glasgow Coma Scale (GCS; Teasdale & Jennett, 1974) is the most widely used quantitative measure of the duration and depth of impaired consciousness following head injury. The benefits of international acceptance, familiarity, ease of use, and a high degree of interrater reliability, weigh the balance very heavily against the introduction of alternative methods of assessing level of consciousness. This is despite the scale's limitations (Marion & Carlier, 1994; Richardson, 1990), which are well-recognized by its originators. Jennett (1989) observed that the GCS

> was not intended as a means of distinguishing among different types of milder injury. Many of these patients are orientated by the time they are first assessed and therefore score at the top of the Glasgow Scale. Yet some of these patients have had a period of altered consciousness, either witnessed or evidenced by their being amnesic for events immediately following injury. (p. 24)

---

[7]A useful method is a brief semantic differential probe that client and caregiver complete independently. Clients' insight into their problems is reflected as the difference between the sums of the two ratings; client and caregiver burden can be separately assessed by scoring the negative weighting of the bipolar adjective pairs (Jansen, 1988).

Serious problems thus arise in applying this scale to the assessment of mild brain injury. Romer et al. (1995) noted that in live hospital admissions, "for every 8 mild cases (GCS 13–15), there is one moderate (GCS 9–12) and 1 severe (GCS less than 9) traumatic brain injury" (p. 14). The problem is that there are 6 scale points at the severe end of the scale (3–8) to accommodate some 10% of cases, and 4 in the moderate range for a further 10%; only 3 scale points are available to accommodate the remaining 80%.

The sometimes severe consequences to which mild brain injuries may give rise are widely recognized by neuropsychologists, but not by other health care professionals. In consequence, access to counseling and rehabilitation becomes more difficult or is denied to clients with an admitting GCS in the mild range. In the developing countries, in which there is especially high reliance on the GCS and the medical diagnosis, which takes little account of the changes noted earlier, an admitting GCS of 14 or 15/15 is often the kiss of death for compensation claims. Even if the victim is unable to return to open-market employment, compensation may be based only on orthopaedic injuries, leaving out of account the disruption of family life and the lifetime loss of earnings that may arise from a mild concussion.

With the support of the World Health Organization Advisory Group on the Prevention and Treatment of Neurotrauma, I have attempted to address this problem by devising an extended version of the Glasgow Coma Scale that defines a set of behavioral landmarks that fix the duration of posttraumatic amnesia, and on this basis codes an additional digit that follows the GCS score. The rationale is given in Nell and Yates (1998) and the method in Nell (1997b).

*Chapter* 8

# Realism and Intensity in the Diagnostic Interview

With all clients to whom testing is a new and unfamiliar experience, and especially in developing country settings, examiners need to be acutely aware of the near-mythical power that clients attribute to professionals, and their fear that the decisions made by professionals will irreversibly affect their destiny. In countries that have had a history of customary racism, unfortunate resonances are set up if (as often happens) the tester is White and the client is Black.

## X-RAY EYES AND OTHER SOURCES OF ANXIETY

Our neuropsychology research group's test manual (Nell & T. Taylor, 1992), prepared for use with farm laborers in South Africa's Western Cape province, begins with this somber warning:

> Although race relations on South African farms are better than they used to be, there are still very strong white–black power relations. A white examiner testing a black subject creates an unwilling replication of these relationships in a literal sense; a black tester, though ostensibly at a lesser social distance, is also in a powerful position and can also set up a master–servant relationship that inhibits responding and depresses performance level. (p. 10)

This bears reflection: Although we live in a post-Freudian age, transference is real, and is powerfully at work in every test situation. The manual continues:

> Anxiety caused by any of these factors—fear of the test situation, concerns that test performance may endanger employment, fear of the examiner's "X-ray eyes," or simply an undue sense of awe resulting from an authoritarian test situation—will undermine your clients' comprehension of test instructions, make them fearful of errors and therefore slower to respond, and affect scores in many other ways. (p. 10)

But there are no short cuts to setting clients at ease. The manual goes on to caution that

> the formality inherent in the test situation must be respected. The efforts of a well-meaning examiner to turn the examination into a social occasion, with smiling, jocularity and the use of first names, will backfire. There is a social distance between examiner and subject that must be respected, and fears must be addressed and allayed in this context. . . . The examiner must . . . greet the client with appropriate seriousness. In languages that have both formal and intimate styles of address [see chap. 7, "Discourse Register"] the examiner does well to retain the formal register, at least for a while.
>
> *Hello Mr Mars. My name is John Bloggs. We're going to be working together here this morning/afternoon.* (p. 11)

This degree of respect and circumspection may not come easily to psychologists, who have traditionally been among the less well-mannered social scientists, unpacking their carpet-bag of tests in a stranger's presence without waiting for the teacups to cool. Instead, like anthropologists and oral historians, we need to join with our clients' world before inviting them into ours.

There are countless other ways in which the neuropsychologist—and test administrators working under the neuropsychologist's supervision—can set clients at ease. With clients unaccustomed to testing, the most important element is unforced warmth in its simplest and most direct manifestations. Sometimes clients will have traveled many hours to reach you. Check if they are hungry, thirsty, or need the restroom. Then preview how long the session will last, and whether the clients are able to stay that long. Communication may not be as good as you would have hoped, and anxious clients may be surprised to learn of the length of the session. If they are worried about the last bus home, they will not do very well—especially if young children are present, or elderly family members have come along in response to your request that the client be accompanied by caregivers. Then explain who you are, why the interview is necessary, why it will take so much time, and, prior to testing, what this involves and how long it will take. Fuller explanations will follow later in the test room, as required by the procedure described in chapter 10. In adversarial matters, you need to take special care that clients understand your relationship to them, their lawyer, and the court.

## THE INTERVIEW AS DRAMA: REALISM AND INTENSITY

The diagnosistic interview is a real-life drama. The hero or heroine is your client, and there are leading roles for spouse, parents, neighbors, and children. The stage set is not your consulting room but your client's home. Later, set changes will make it the workplace or the school. You are the director: You set the scene and decide the sequence, sometimes letting the dialogue flow, sometimes intervening to maintain intensity. The actors know their lines; they live them every day. But unless you, as director, maintain the intensity, your consulting room stage will not reflect real life, but show bored people talking about distant things that happen somewhere else, rather than here and now. When people are bored, they want to get away as soon as they can: Your clients will tell you more, and you will learn more that is diagnostically useful, if your dramatis personae are charged with energy, reliving their home in your office. The drama is diagnostic.

### Dramatis Personae

In setting up the appointment, you need to specify who should accompany the client. Generally, these are the adults who live under the same roof as clients and spend most time with them. For adult clients, the prime informant is the cohabiting companion; next are the parents, and then close friends (i.e., on condition that they knew the client well before the accident and have maintained contact since then so that they are able to speak to whatever changes have taken place). If the client is a child, then you may find that grandparents living with the child or nearby are better informants than the parents. You cannot take it for granted that the mother is best informed; the father may spend longer hours with the child. Siblings can fill in a lot of information, but should usually be interviewed separately. For both adults and children, neighbors may be surprisingly useful. In squatter settlements, increasingly a feature of life around developing country cities, shacks look into one another, and neighbors sometimes know more than family members.

Prime informants will not be prime at all if they are poor observers. I have had university-educated mothers who are devoted to their children say that little Melanie's memory is fine and she does not forget things any more than her very bright siblings—yet on testing, her memory is devastated. "Ah, yes," says mother thoughtfully, "Now that you mention it I suppose you're right. We do have to get her homework from the teacher ourselves, and I always pack her bag in the morning so she won't leave things behind"—and this despite 20 minutes of probing about memory during the interview! The quality of your informants is something you have to feel out during the interview. But, you dare not base your behavioral description on an informant who just doesn't perceive behavior.

## The Cinderella of the Health Sciences

Ideally, the neuropsychologist-examiner should be fluent in the client's home language. Under conditions of mass migration across language borders, in Africa and eastern Europe, and in settings of linguistic diversity, this is often impossible. India has 16 official languages and 24 that are spoken by more than a million people; there are 12 official languages in South Africa. No single translator can be fluent across this range of languages and dialects.

What is little appreciated is that under these circumstances, assessment quality is determined by translation quality! Yet clinical translation remains the Cinderella of the health sciences. In *Wretched of the Earth* (1968), Franz Fanon wrote that the inhumanity of colonial medicine as he witnessed its practice in a psychiatric hospital in French Algeria in the 1950s was compounded by the language barrier. To get a patient history, the French-speaking doctor would waylay the first Arab speaker he could find, and this porter, cleaner, or nurse aide would convey the patient's complaints to the doctor in broken French. It is inconceivable that psychiatric assessments could have been carried out under these conditions of linguistic barrenness and inaccuracy, but such travesties of patient intake continue in many settings. In neuropsychology, such practices are intolerable because subtle complaints are lost if they are mangled in translation by an attorney's driver or clerk. It is even worse to use a member of the client's family as a translator or test administrator. Moreover, because neuropsychology translators are necessarily also test administrators, they should have at least a 4-year psychology degree followed by substantial supervised experience.

## Reflection as a Translation Technique

One of the difficulties that arises even with the best translation is that the client and caregivers are soon marginalized. The translator interacts with the neuropsychologist in a language the clients do not understand; in fact, the neuropsychologist seems to the clients to have no interest in them, because after she has asked a question she begins writing, pays no attention while the family is speaking, and listens again only when the translator speaks to her.

An innovative remedy for this unfortunate state of affairs is to use the therapeutic modality of reflection as a translation technique. The way this works is that your question or comment is translated to the family. Then, instead of speaking the translation to you, the translator repeats the answer, but now in English, to the speaker. You will notice that in the exchange that follows, the client replies directly to the neuropsychologist's question "How bad does that get?" with no intervening translation. This is because many clients have good receptive skills in the language of translation despite being unable to express themselves fluently. But, if you address them directly, they may respond to you directly, although in their own language—a more immediate state of affairs that avoids one leg of the translation. The following example illustrates the use of reflection:

Translator: Do you feel you have changed since the accident?

Client: My wife makes me angry now. Before I was sometimes angry, now I am angry all the time, for little things, for nonsense.

Translator (maintains eye contact with the client; this in turn compels the neuropsychologist to focus on the family rather than the translator): You're saying that you used to get cross with your wife sometimes, but now you are angry with her all the time. This anger is caused by unimportant things.

Neuropsychologist (keeping eye contact with the client as she speaks): How bad does that get? When you are angry, do you talk loudly or do you shout? Tell me what you do.

Client: Yesterday I took this crutch and I tried to hit her, I was so angry. Then she ran away. Before, last month, I caught her and I wanted to hit her, but my son came, this one here, and he held me fast.

Translator (maintains eye contact with client as before, as does the neuropsychologist): You say that you sometimes get so angry . . .

Whether or not the technique of reflection is used, a useful aid for the immediacy and acccuracy of translation is to teach translators to render both the client's and the neuropsychologist's speech in the first person, not converting it to indirect speech, to translate fully, and to translate frequently. An old South African psychiatric hospital joke concerns a client, a psychiatrist, and a nurse-translator. "Are you hearing voices?" asks the psychiatrist. An animated conversation ensues, with an excited client gesticulating and the nurse listening animatedly. After 10 minutes, the nurse turns to the psychiatrist and says, "He says 'Yes,' doctor."

### Moving Between Client and Caregivers

The imaginary client in the previous vignette is telling the truth as he sees it. How do you get at other views of "the truth" without slighting the client? At the beginning of the interview, explain that people very often see the same things in different ways, and that you will be putting the same question first to the client and then to the caregivers "not in order to check up on you, but to see how other people feel about the same thing." Swinging between client and caregivers is an essential aspect of the diagnostic interview; the danger is that it can also become boring unless neuropsychologist and translator keep up a cracking pace to maintain the energy level in the room.

The emerging pattern of complaints will give you a good idea of where stress is heaviest in the family. A good principle to follow, especially if the assessment will be followed by therapy, is that there are no secrets, and everything will be discussed openly in the presence of the family and the client. However, under some circumstances, this is just not possible and the interview will have to be split (e.g., with an angry client and frightened caregivers).

### Collateral Information

It will often be necessary to interview teachers or work colleagues to amplify the picture provided by family caregivers. A trap lies in wait here: If you interview the most senior person in the organization—school headmaster, police chief, or production manager—you will get an arm's length evaluation that has been filtered down from the primary sources of information and couched in bureau-speak, which is a diagnostically useless language of circumlocutions. The trick is to get to the primary sources—but that is not easy, because on sensitive matters, the senior person is often the only authorized spokesperson. But the people who will give you the information you need are the class teacher who spends 6 hours a day with your client, the buddy who rides with him in the police car, or the workmates whose life he puts at risk on the production line. But, because he is their friend, your client's workmates will lie to the boss about his performance because, if the boss knew the truth, he would fire him. This is touching, but can cause the client huge financial loss by blocking the compensation to which he is entitled. Consider the following case:

> A 42-year-old foreman at a precious metal refinery had returned to work 18 months previously after a severe head injury. Performance evaluations in the client's personnel file were positive, with both the production manager and works supervisor recommending promotion to shift boss. On the other

hand, the behavioural interview—confirmed by the test results—indicated that gross impulsivity, poor judgment, unreliable memory, and temper dyscontrol were present. The conundrum was resolved when a shrewd industrial psychologist visited the plant. He was given permission to inspect the client's job site—a walkway above a slurry drying plant that operated at a temperature of 2000 degrees centigrade—and to talk with the crew that worked daily with him. Once these workmates understood the benefits their frankness would bring to the client, they said that he had on several occasions caused dangerous situations, and that they had in effect taken over his foreman's duties, leaving him with only routine tasks at some distance from the furnace. They confirmed that in order to protect their friend, they had made misleading reports about his job performance to plant management.

## SYSTEMATIC INTERVIEWING

Most neuropsychological assessments of diffuse traumatic brain injuries take place a year or more postaccident; their purpose can be simply stated. It is to determine what clients would have achieved but for the accident, and what they are now likely to achieve, taking the accident into account, and making allowance for the remediation you recommend. "But for the accident" is an evocative term, projecting the neuropsychologist into a speculative inquiry that is historically based. It is founded on detailed information about the client's family background, home circumstances, and school and career achievements. These questions determine the structure of the intake interview.

But unless this interview is carefully structured, the clients' press to talk about what is most important to them (this will often be a description of the accident or hospital treatment) will suck the interviewer into extensive note-taking that later proves useless in compiling the report, while failing to get vital information such as educational background, achievements of parents and siblings, and employment history. However, the worst way to conduct an interview is by working doggedly through a checklist. Each client is different, but a checklist forces the interviewer to gloss over the unique and deal at length with the unremarkable. Also, checklist interviewing becomes stupefyingly boring for both the interviewer and the client.

## INTAKE FORM

On the other hand, a standardized intake form allows the examiner to pick up on the unique parameters of each case in a time-economical way. In part, this is because the intake form takes care of the routine aspects of the interview ahead of time, and in part because after some experience, you conduct the intake interview by scanning the form as you dialogue with the clients, glossing over the usual and expected; the client completed 10 years of school, worked as a scaffold builder and carpenter for a construction company, then—aha!—worked for the next 3 years for a TV production unit as a set builder. "Won't you tell me about that," you ask, sensing upward employment mobility that says something about the client's energy and ambition, and if this is a forensic assessment, affects the compensation entitlement. And so goes the interview, focusing on the hints the form gives of significant aspects of the history and current behavior.

Each setting will develop its own intake form geared to its particular client profile and documentation needs. As a guide to coverage and format, the version our practice uses is given in Appendix 2. For literate clients, the form is self-administering and should be mailed ahead for completion prior to the interview. For semiliterate clients, it is completed by the examiner at the initial interview.

### Intellectual and Social Endowment

The questions in this section, together with the educational and occupational history, give you most of the material you need to describe the accident scenario. You will see that Section 2 of the form elicits information about the education and occupations of parents and siblings (if time permits, it is also helpful to get educational and occupational information about the grandparents on both sides); and, whatever one's views about the relative contributions of heredity and the environment to individual development, it is essential to have this information. With regard to level of education, both for the client and for the client's parents, we have found that highest standard passed correlates better with other performance indicators than total years of schooling.

The career level and place of residence of the parents give an idea of the stimulus richness of the home and neighborhood environment. This information should (as noted in chap. 5) be supplemented by questions about the socioeconomic status of the home (e.g., the availability of books, magazines, radio, television, etc.). For children who sustained brain injuries early in their schooling, the family background is often the only way of determining premorbid ability.

### School Record

The most direct way of gauging injury impact on schooling is to construct a year-by-year table setting out grades, your client's average as against the class average, and the client's rank position in the class. However, in some school systems, classes are so large and teachers so overworked that class grades are inaccurately assigned; teachers' written comments in the school report may be high on goodwill but low on pertinence to the child in question. In these circumstances, the most useful question to ask children is where they place in the class. Sometimes there is a formal ranking that may be reflected in the report; if not, children often informally rank themselves, and can tell you with some accuracy whether they place near the top or bottom of their class, and with some probing, give their position to within a few places. It often happens that child and mother remember where the child placed before the accident, and can compare this to present achievements, allowing you to quantify injury effects. A particularly useful item of information is classes failed, when this happened in relation to the accident, and the reasons the client and caregivers give for the failure. "She was lazy" or "The teachers were on strike" (a twice-a-week occurrence in South Africa in the 1980s and 1990s) is one thing, and "She couldn't understand" is another.

Often, parents have had direct contact with teachers. You need to ask if there have been any complaints about the child's behavior at school. These will often relate to arousal and disinhibition: The child may concentrate poorly, losing the thread of the lesson and giving off-the-point answers when called on, or actually fall asleep in class. Slowed information processing leads to complaints that the child cannot keep up with the class. Memory problems reflect informally in missed homework assignments and failure to pass on messages from teachers. Reduced inhibitory control results in disruptive classroom behavior (e.g., disrespect to the teacher, talking loudly during the lesson, or altercations with other children). Another important behavioral indicator is playground conduct, but in overcrowded schools this information is hard to come by. Teachers do not know unless there has been a major incident; if this information is needed, then it may be useful to locate a present or past friend in the same class.

For adults, information about postschool training is important for the light it throws both on intellectual capacity and drive.

### Occupational History

For adults whose schooling is well behind them, education remains a primary determinant of test scores, but the occupational history has precedence in answering the "but for the accident" question. For working-class clients whose job descriptions may not change much, upward mobility will be most clearly evident in the earnings record. In the developing countries, there are startling opportunities in the informal sector (e.g., running a hairdressing or grocery store from home, roadside hawking, door-to-door trinket sales) and clients' achievements in such enterprises must be fully recorded. One thing the very rich and the very poor have in common is the lack of formal income records that can stand up to court scrutiny. High-earning restauranteurs and shopowners squirrel away all the cash they can so as to pay as little tax as possible; similarly, informal sector entrepreneurs cannot prove their earnings. The following is one such case:

> A young man with a strange gift came from the rural north of South Africa to Johannesburg: he could neither read nor write and spoke no English, but prospered as a seer who could read the future in the flame of a candle set on the table between him and his client. He earned the equivalent of $100 dollars during the week, and an additional $200 on weekends. Six years later, having brought his wife and child to the city, he was caught in crossfire between the police and a gunman, sustaining a right frontal bullet wound. He was referred for assessment by his attorney and presented as a moderately depressed, mild-mannered, and singularly charismatic individual. He had only one complaint—that since his injury, he could no longer read the candles. His loss of income was very substantial but could not be proved. The police insurers paid out a derisory sum. His only income thereafter was the monthly state disability grant, equivalent to about $75 a month.

### Westernization and Literacy

At about this point in the interview you have the information you need to decide which core battery tests you will be using with your client when the interview ends. The parameters for this decision are based on what you have learned to this point about your client's home circumstances, level of urbanization and education, occupational history, and home language: As A. J. van den Berg at the South African Human Sciences Research Council has noted (personal communication), translating a test from English into the clients' vernacular may block access to terms and concepts they have acquired through English.

## THE BRAIN DAMAGE INCIDENT AND RELATED EVENTS

You will notice that on the intake form the questions about the incident, hospitalization, and amnesia appear as Section 8, the last in the form. This is because these questions are typically the most puzzling for clients, and if they appear before the "Main Complaints" section, where they belong, some clients give up in despair and arrive with this section blank, and everything after it. However, at the interview, it is important to deal with the events around the injury at this point so that you can relate outcome to severity.

### Grading Injury Severity

From prior reading of the documentation, you will have some idea of initial injury severity. As previously noted, the Glasgow Coma Scale (GCS) is the most widely used measure of the depth and duration of unconsciousness, but is relatively insensitive to mild brain injuries, which account for 80% of all live hospital admissions. The Glasgow Coma Scale (Extended) (GCS–E; Nell, 1997b) described in chapter 7 promises to enhance the upper level sensitivity of the the tradional GCS, without undermining the latter's proven utility.

Another important indicator of injury severity is posttraumatic amnesia. For clients who do not have the gift of introspection (the return of continuous memory is an abstract and difficult notion), this can easily turn into the most confusing and protracted part of the interview. Some of these difficulties are avoided by the direct and concrete questions in Section 8. Your client's answers to these, interpreted against the review of posttraumatic amnesia in chapter 7, will more easily allow the determination of PTA.

Question 8.1 provides for an estimate of the duration of retrograde amnesia, because the circumstances of the last memory can usually be brought into a temporal relation with the accident. Even if PTA is prolonged, retrograde amnesia is typically of brief duration (i.e., a few minutes to an hour or two), although it can exceptionally extend to a year or more. Very rarely, it may be total. Consider an example:

> A young woman sustained a devastating pedestrian head injury at the age of 19 which left her comatose for six weeks and with a prolonged post-traumatic amnesia. A year post-accident, it became clear that her entire past history had been lost. Despite detailed probing, she had no recall for the Christmas a year and two years before, nor for her school-leaving dance, her classmates or class teachers in high or junior school, her first day at kindergarten, or any other childhood event. When seen for assessment at three years post-accident, she said that with the help of her parents, sister, and friends, she had reconstructed some past events—though this was a tenuous reconstruction, since her recall of new information was poor. The effect of the total retrograde amnesia in combination with the amnesic deficit was that this woman lived in an evanescent bubble of present time.

***Previous CNS Compromise.*** Tucked away at the end of the intake form (Item 8.12) is a question about previous episodes of loss of consciousness, worded in such a way as to include nontraumatic events such as a partial drowning, being overcome by carbon monoxide, domestic gas, or noxious chemical fumes, damage arising from anesthesia, as well as other trauma such as becoming dazed or unconscious from a fall, a fight, an assault, and so on. To prevent misattribution of old damage to the recent incident, it is imperative to determine whether prior CNS compromise exists: This is especially important in developing country settings. In Western countries, the base rate of traumatic brain injury is relatively low even in high risk groups; in South Africa, which may not be untypical of other developing countries, the rate is higher overall than in the U.S., and very high in high risk groups, for example 763 per 100,000 p.a. for African males from age 25 to 44, and 986 for Colored (i.e., South Africans of mixed descent) males from age 15 to 24 (Nell & Brown, 1991). In South Africa's Western Cape Province, for the Colored farm worker group reported in chapter 2, 69% of the 243 subjects reported a previous episode of dizziness or unconsciousness, of whom 24 subjects (32%) had been dizzy or unconscious three times, 12 on four occasions, and 1 on five occasions (Nell, Kruger, T. R. Taylor, Myers, & London, 1995).

## MAIN COMPLAINTS

As all therapists know, spontaneous complaints have greater diagnostic weight than those that are elicited by direct questioning. In turning now to Section 7 of the intake form, you put a broad, open-ended question first to the client and then to the caregivers: "Have you changed as a result of the accident? What is different now?"

### Spontaneous Complaints, Elicited Complaints, and Probing Without Leading

The spontaneous complaints by client and caregivers seldom exhaust the full spectrum of problems, many of which are subtle and remain unarticulated. It is therefore necessary to move beyond the information you have been given and to probe to get the additional information you need about other changes you suspect may be present—in arousal, alcohol tolerance, temper, self-monitoring, finishing tasks, goals and plans (which are two very different things!), memory, and problem solving.

Probing is one thing, but it is quite another to ask leading questions that will get you the answers that consciously or unconsciously you want to hear. "Do you find that you sleep more now than before?" is a leading question. "Have you noticed any change in your sleeping habits since the accident?" is not. In adversarial matters, experts who allow their bias to suggest leading questions do themselves, the client, and the court a disservice; in therapy-oriented assessments, the temptation is not bias but to save time. Experts know what the most likely answer to a question about concentration or bad temper will be, and they get to be "more efficient" by saying, "Do you find you get angry more quickly now?" Again, this is a short cut that is a disservice to the client.

Once the spontaneous complaints are exhausted, your quest for additional diagnostic clarity begins. However, in order to probe without leading, you must begin with what your clients have brought to the session. So you start by getting more information about what has already been recorded on the intake form or said: "Mr. Mohale, you told us just now that your son is no longer respectful to you. I'm not sure what that means. Can you give me an example?" You will also pick up complaints on the intake form that need elaboration. For example, there may be a statement, "breaks dishes." Is this clumsiness arising from a sensory or motor impairment, or is it the result of carelessness that may point to a self-regulatory problem, or a manifestation of bad temper? You need to find out.

Both in this phase of elaboration on information given or hinted at, and in the next, when you probe for additional information, pay special attention to the classic dimensions of the concussion syndrome as outlined in the previous chapter—changes in arousal, in personality, and in thinking. Familiarity with the cardinal manifestations of traumatic brain injury is the basis for this further systematic exploration of the client's functioning: The material in chapter 7 should not be used as a checklist, but as a guide to the formation of informed hypotheses about where further questioning is likely to be useful.

## Physical Complaints

Physical difficulties are concrete, easily described, and the kind of thing that one is supposed to tell a doctor. Thus, in response to your open-ended question about what has changed, clients will almost invariably begin with complaints about headaches, the arm that is weak, and the leg that hurts when they walk. However, some physical impairments are puzzling and may not be volunteered: There is no need to work through all the possibilities in checklist style, but with regard to the subtler deficits reviewed in chapter 7 (i.e., olfaction, sensory integrity, visual field cuts, etc.), it is necessary to ask about any areas that you suspect may have been compromised on the basis of the information you have gathered to that point.

As noted in chapter 7, it is important that physical difficulties are documented. Indeed, it is sobering to realize that a complete neuropsychological test battery can be administered without determining whether your client can walk unaided, has full use of both hands, or can see well enough to read or watch TV. In some settings, neuropsychologists routinely test visual fields, visual and auditory extinctions, and peripheral sensation. Where psychology is a young or weak profession, such examinations could call down the wrath of the medical establishment. But you can still ask your client to walk across the consulting room, and to read fine print figures or numbers aloud.

However, this phase of the interview can easily run away with you. Clients' natural desire to talk at length about their injuries, hospital treatment, and medical consultations must be contained.

***Chronic Pain.*** A brief digression on the subject of pain is required. Just as translation is the Cinderella of interviewing skills, chronic pain is the neglected aspect of the diagnostic interview. Victims of traumatic brain injury have more often than not also sustained polytrauma, which may have left them with severe residual pain. However, pain complaints are often glossed over by neuropsychologists and other experts as being outside their field or, worse, the manifestation of "hysteria" (in the MMPI rather than the Freudian sense—although both may apply!). A chronic pain workup by an expert in this area (often an anesthetist or a specialized health psychologist) can open the door to appropriate therapy for a client who has not hitherto been taken seriously.

## CHANGES IN PERSONALITY AND INTELLECT

Having dealt with the physical sequelae, you say "Apart from the physical problems we have been talking about, do you feel you have changed in other ways?" Often, this question is like an electric current running through the room, and many of those present poise themselves to speak. Precedence, as always, is given to the client, and thereafter to others present. Here are two hints: At this stage, the less you say, the better, both to preserve your neutrality and to give free rein to the family. Second, bear in mind that in some traditional cultures, age has precedence, and women may be reluctant to speak before men. Whatever one's own views, these norms should be respected.

However, the electric current may not be generated. Some clients and caregivers are poor informants, and it may be necessary—without compromising your neutrality—to canvass the dimensions of personality change reviewed in chapter 7 that you think may be relevant.

Chapter 9

# Buds, Flowers, Fruits: Potential, Performance, and Test Administration

In a 1926[1] paper, Vygotsky (chap. 3) broke with three central tenets of the psychology of his time: that IQ is immutable, that learning necessarily trails behind development, and that if children's mental functions have not matured to the extent that they are capable of learning a particular subject, then no instruction will prove useful. "Development or maturation is viewed as a precondition of learning, but never the result of it" (1988, p. 80). Vygotsky went on to propose a radically new approach that distinguishes between the actual developmental level of the child, and the level determined by intellectual testing.

The testing looks backward, argued Vygotsky, at development that has already been completed: This means that if children barely miss an independent solution of a problem—for example, if they solve the problem after the teacher has initiated the solution, or after leading questions have been asked that indicate how the problem might be solved—then the solution is not regarded as part of the child's mental development (p. 85). On the contrary, commented Vygotsky: "What children can do with the assistance of others might be in some sense even more indicative of their mental development than what they can do alone" (1988, p. 80).

Vygotsky then defined his central developmental construct, the zone of proximal development (the ZOPED), as the distance between the child's actual developmental level during independent problem solving, and the child's level of potential development revealed by problem solving under adult guidance or when working with more competent members of the child's own age group. This notion of "distance" at once turns the exam-

[1]Reprinted as chapter 6 of *Mind in Society* (1988a).

iner's attention from performance to potential, and from being to becoming. In a famous passage, he wrote:

> The zone of proximal development defines those functions that have not yet matured but are in the process of maturation, functions that will mature tomorrow that are currently in an embryonic state. These functions could be termed the "buds" or "flowers" of development rather than the "fruits" of development. (1988, p. 86)

This has been a fertile idea in modern psychology. Cicerone and Tupper (1986) proposed an adaptation of the ZOPED as a "Zone of Rehabilitation Potential" that has special relevance for metacognitive interventions with brain-injured persons. The major contribution to the assessment of learning potential has been made by Reuven Feuerstein (Feuerstein, 1979, 1980; Feuerstein, Rand, Hoffman, & Miller, 1979), who adopted Vygotsky's notion of mediated learning, but, in a striking omission, did not cite Vygotsky's work. Feuerstein argued that cognition can be modified, especially in apparently "retarded" adolescents who have developed in the context of sociocultural deprivation. Such deprivation results in a lack of "mediated learning experiences" (Feuerstein, 1979, p. 539), which occurs when environmental events are "invested by specific meaning by mediating agents" like parents or teachers. Through such mediated learning, continued Feuerstein, children acquire cognitive schemes through which their repertoire of intellectual functions is constantly modified.

The cognitive impairments that arise through such deprivation are deficiencies at the *input phase* that affect the quantity and quality of input data and arise through such problems as blurred and sweeping perception, impulsive and unsystematic exploratory behavior, impaired spatial and temporal concepts, imprecise and inaccurate data gathering, and the like. At the *elaborational phase,* the deficiencies impede efficient use of available data, for example, inadequate problem definition, difficulty in selecting relevant cues and disregarding the irrelevant, impaired strategies for hypothesis testing, and impaired planning. *Output phase* deficiencies give rise to inadequate communication of solutions and arise because of egocentric communication, trial and error responses, poor verbal skills, imprecise responses, and so on (Feuerstein, 1979, pp. 61–87). The parallels between "cultural deprivation" as described by Feuerstein and the consequences of neuropsychological impairment are striking.

## GUIDED LEARNING IN SOUTH AFRICA

Despite sporadic exceptions (the most welcome recent example is the WAIS–III; see chap. 10), psychometric testing has remained an arm's-length procedure: In the interests of standard test administration, examiners must read from the manual, not changing the wording in any way; directions may be repeated but not explained; and the client can be given no indication of whether an answer is right or wrong. The standard answer to clients' puzzled questions about what is expected of them is, "Whatever you think is best," which is a response that distances the examiner from the client and establishes an unbridgeable professional distance between them.

It is thus to the credit of many South African psychologists that during the heyday of this hands-off approach, they were able to say, as Biesheuvel did in his directions for the South African General Adaptability Battery in 1950, that tests should involve a learning component, and that "the subject should have ample opportunity to learn to do the task involved in the test by preliminary exercises. . . . The test should also provide scope for the insights . . . and experience acquired in the course of testing to progressively improve performance" (Biesheuvel, 1972). Accordingly, most of the tests provide for a number of trials, or for tasks of increasing difficulty so that subjects have an opportunity to try different strategies and become aware of their errors.

The intensive coaching Crawford-Nutt (1976; see chap. 4) devised for Black subjects on Raven's Progressive Matrices, leading to striking improvements in their scores, is very close to the kind of guided instruction and collaborative problem solving that takes place in the zone of proximal development; the same can be said of M. A. Verster's findings on the consistent improvement by Black mineworkers on repeated exposures to the Classification Test Battery (chap. 4). Although testing conditions did not change from one administration to another, and there were no client–examiner interactions, the experience of repeated exposure itself constituted a ZOPED within which these clients, repeatedly observing the demonstration film, reflecting on their own performance and the work methods of other miners in the room, were able to acquire increasing test-wiseness. Indeed, the performance gains Verster observed appear to be of the same kind as the triggering effects found after training on piagetian tasks in West Africa (Dasen et al., 1979), and in Ciskei, South Africa (Rogan & MacDonald, 1983).

## LEARNING POTENTIAL AS A MEASURE OF ABILITY

Since guided learning experiences do indeed produce significant performance increases, the distance between actual performance, and performance after triggering in the zone of proximal development, might be a more reliable guide to an individual's capacity to benefit from training than a simple performance measure that is applied indiscriminately to test-wise clients, and those from socially and culturally different backgrounds. Accordingly, one of the most intriguing challenges in modern psychometrics is to develop a measure of learning potential that has good predictive validity. But since the 1950s, this has remained tantalizingly out of reach. For example, Dague (1972) cited the work of Ombredane and his colleagues in the 1950s in which the gains on repeated test administration were taken to measure learning potential among young Congolese in the 10- to 14-year-old age group. But far from predicting educability, correlations of educational success with difference scores were poor. Shochet (1986, cited in Boeyens, 1989) adapted Feuerstein's method to investigate learning potential among disadvantaged first-year bachelor of arts students at the University of the Witwatersrand in Johannesburg, South Africa. Test gains from pretest through to a posttest following intensive coaching failed to correlate with any measures of academic performance. Boeyens (1989) noted that this and many other studies of learning potential "evidence a complete disregard for the problems of reliability" (p. 40); for Shochet's study, the mean difference score was so low as to be uninterpretable. T. R. Taylor (1994) distinguished between Type 1 potential, achieved through the teaching of thinking skills, and Type 2 potential, measured by tests of learning, and speculated that Type 1 potential, measured as test–retest gains on novel tasks, may eliminate differences that arise from test-wiseness and culture.

## EDUCATING THE EXECUTIVE

A theoretical note that articulates Vygotsky's ZOPED with Sternberg's information-processing paradigm (chap. 4) will be helpful at this point. Although all three aspects of Sternberg's componential theory are affected by environment and experience, it is the executive itself—the "little person" up in the brain—that decides whether to emphasize speed or deliberateness, what to automate and what to leave to the global processor. This is because the metacognitive "drivers" of the processing system are most heavily governed by environmental press and cultural values (Verster, 1986).

To the extent that executives are mental representations of tasks and behaviors determined by the culture, they are accessible to consciousness and are able to give some account of themselves in conversational inquiry. It therefore follows that by using the method of extended conversational inquiry described in Appendix 3 in relation to the Digit Symbol subtest of the WAIS–III, it is possible to determine whether the client's metacognitive drivers have constructed the given task in ways that are equivalent to those in the examiner's mind.

## BRINGING THE ZONE OF PROXIMAL DEVELOPMENT INTO TESTING

The computer-based Performance Probe Test Battery (Nell & T. Taylor, 1992), developed for our group's study of Western Cape farmworkers, and our unit's *Core Battery of Psychological and Neuropsychological Tests for Persons with Less Than 12 Years of Education* (Nell & Maboea, 1997) for a range of paper-and-pencil tests, set out to bring the zone of proximal development into the test administration procedure within the parameters of stable test administration (chap. 10). This is done through an extended practice procedure that incorporates a guided learning experience. The practical application of this method is now illustrated by first describing the Hick Box administration of the Simple Reaction Time task, and then for the Rey Complex Figure Test. It should be noted that although referenced to these specific tests, this method is in broad outline applicable to all psychological tests. Thus, the Hick Box method is *mutatis mutandis* applicable also to the Terry Reaction Time Device used in the WHO–NCTB (for details of this adaptation, see Nell & Maboea, 1997).

## EXTENDED PRACTICE FOR AN APPARATUS-BASED ITEM (HICK BOX)

As with all the other tests in the computer-administered Performance Probe Battery, administration of the Hick Box reaction time test has five phases:

1. Apparatus familiarization.
2. Instructions in a slow question-and-answer format. Because this subject group was rural and functionally illiterate, concrete language and vivid examples were used. Readers should bear in mind that although the farms are heavily mechanized, the milieu in which the test subjects live is closer to the situation in Luria's Uzbekhistan and Gilbert's rural Kwa Zulu (chap. 3) than to a modern urban environment.
3. Thirty practice trials. On the tests using the computer screen, only the correct response key is live, allowing the examiner to discuss with the subject the reasons for wrong responses. If the subject seems confused or hesitant at the end of this phase, then it is repeated as a second guided learning loop. If after this second practice administration the examiner is convinced that further instruction would be fruitless, then the test in question is abandoned and another tried. It has never in our experience been necessary to abort the entire session.
4. Twenty speeded practice trials on which the subject is encouraged to work as quickly as possible without making mistakes. This is an absurd requirement, as pointed out in chapter 1, but at the present time it is unavoidable.
5. The test proper.

To illustrate the coaching procedure, the instructions given in the test manual are quoted verbatim here. These instructions can readily be linked to the concept of weak construct equivalence defined in chapter 5: They have the effect of "educating the executive" by allowing the superordinate metacognitive processes to grasp test demands and rehearse strategies to as great an extent as possible within the limits of a practical testing situation.

### Apparatus Familiarization

*For the next few tests, we are going to be using this box.*

Examiner shows Hick Box without masks.

*The test always begins the same way. You keep the your finger on this button down here at the bottom.*

Examiner places own index finger on the Home Button (HB).

*Show me how you do that. When your finger is on this button, you have to keep it pressed down. Okay. You can take your hand away for the moment. What happens next is that one of these 8 lights will come on. It might be this one . . .*

Examiner presses HB to illuminate Light 2.

*. . . or this one (Light 8) or this one (Light 3) or this one (Light 5). Whenever a light comes on, your job is to switch it off as quickly as possible, like this.*

Light 4 comes on and examiner moves finger from HB to button under Light 4 very quickly.

*You have to do that very quickly, and all you do is press the button under the light that went on. Let's try that a few times. Put your finger on this button now. We call it the Home Button because you always go back to it after you have put out one of the lights. Now whenever you see a light go on, switch it out as quickly as you can.*

Press HB.

*This is like a cowboy movie. That light over there wants to shoot you. You have to put it out so quickly that it doesn't get a chance. It's like a fight between you and the light. You have to shoot that light before it can get you. Now let's try again and see how quickly you can knock out that light. But don't hit the button too hard!*

Examiner demonstrates quick but gentle movement. If the subject does the movement instantaneously and as rapidly as possible, go on to the next phase; if not, repeat the instruction loop.

**Instructions in Slow Question and Answer Format**

The overlay exposing Light 5 (see Fig. 9.1 for explanation) is used. The trial begins with the subject holding down the Home Button with the index finger of the dominant hand. A preparatory beep sounds as HB is depressed, and after a random interval of between 1 and 4 seconds, the exposed light illuminates, and remains on until the subject depresses the button underneath the light. Forty trials are given.

*Okay now we are ready to begin the first part of the test. In this part, we're going to use only one light, so I am going to cover up all the other lights except this one.*

Examiner points to Light 5 and the button underneath it.

*You are going to begin each time just as we practised with your finger on the Home button down here.*

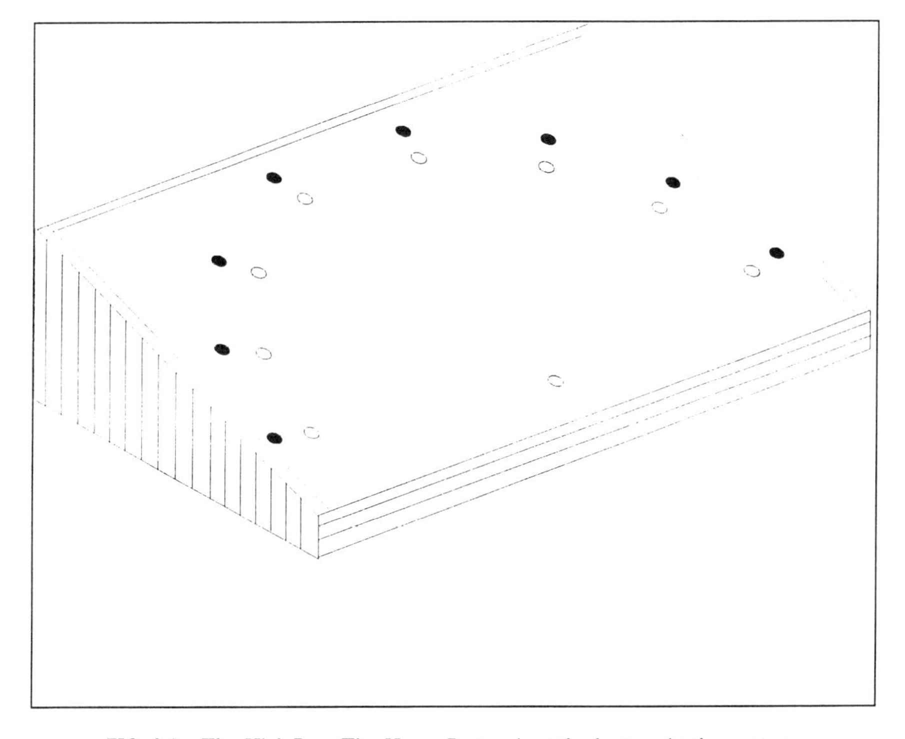

FIG. 9.1. The Hick Box. The Home Button is at the bottom in the center. The 8 lamps (filled circles) run in a semicircle, each with a switch button (open circles) next to it. In the examiner's mind (but not on the Hick Box itself), the lights are numbered clockwise from 1 at the bottom left. For simple reaction time, all the lights except 5, just to the right of the midline, are masked with a wooden overlay.

Examiner demonstrates.

*When this light comes on, all you have to do is switch it off as quickly as possible, like this.*

HB illuminates the single exposed light after a 2-second pause. The Examiner presses this key and says,

*Now watch what I do. When that light [points] comes on, I shoot it dead with my finger very quickly.*

Examiner demonstrates when light illuminates.

*Now we're going to practice. Put your finger on the Home Button. As soon as you see the light come on, put it out as quickly as you can and then come home to the Home Button. But you must wait for this light . . .*

Examiner points to Light 5.

*. . . to come on before you take your finger off the Home Button. If you lift your finger off before the light comes on . . .*

Examiner does so.

*. . . this buzzer sounds . . .*

Examiner points to buzzer and to the screen display, "lift, trial aborted."

*. . . and the test stops. The computer says you lifted your finger too soon, so the test stops and I must start the test again.*

*Now you try that. Put your finger here [HB] and don't lift it till the light comes on.*

When light illuminates, Examiner says,

*Now! Put the light out quickly! Do that again—but this time lift your finger off before the light comes on, now! You see, the buzzer sounds and the test stops. You must always wait for the light to come on before you lift your finger off the Home Button.*

Examiner continues practice trials as previously, repeating practice if necessary. When it is clear that the subject understands the task, the examiner says,

*Right, I can see you know exactly what to do,*

and goes on to the practice phase.

### Practice

*Now you'll work on your own. Are you ready?*

Thirty practice trials follow. Examiner intervenes if the subject makes a mistake, establishes the reason in dialogue with the subject, and encourages him to continue.

### Speeded Practice

*Good! Now you know exactly what you have to do. From now on, I want you to work as quickly as you can without making mistakes. As soon as you see the light go on, press the button under the light. The quicker you work the more points you get for right answers. But you don't get points for mistakes, so you must be as quick as you can without making mistakes. Start now.*

Twenty speeded practice trials follow.

### Test Proper

*Okay, now let's do some more. Again, I want you to work as quickly as possible without making mistakes.*

Note that the examiner does not say, "This is the real test," or "Until now we've been practicing, now we're going to do the proper test." Such remarks achieve nothing except make the subject more anxious.

## EXTENDED PRACTICE FOR A PAPER-AND-PENCIL ITEM (REY COMPLEX FIGURE)

Practice for this item ensures that non-test-wise subjects clearly understand the importance of accuracy in their reproduction of the drawing.

Hand subject a few sheets of unlined paper, and place the practice worksheet (Fig. 9.2) to the subject's left (to the right for left-handed persons). Say:

*I want you to copy this drawing as exactly as you can. You see, it is a complicated shape. This first one is for practice, so try hard to make your drawing exactly like this on here. Make it the same size and the same shape. Work at your own pace—there's no time limit. I'm going to change the pen colour you're using a few times, but don't let that bother you. When I change the pen colour, you just carry straight on from where you are.* (Hand subject a pen). *Please begin.*

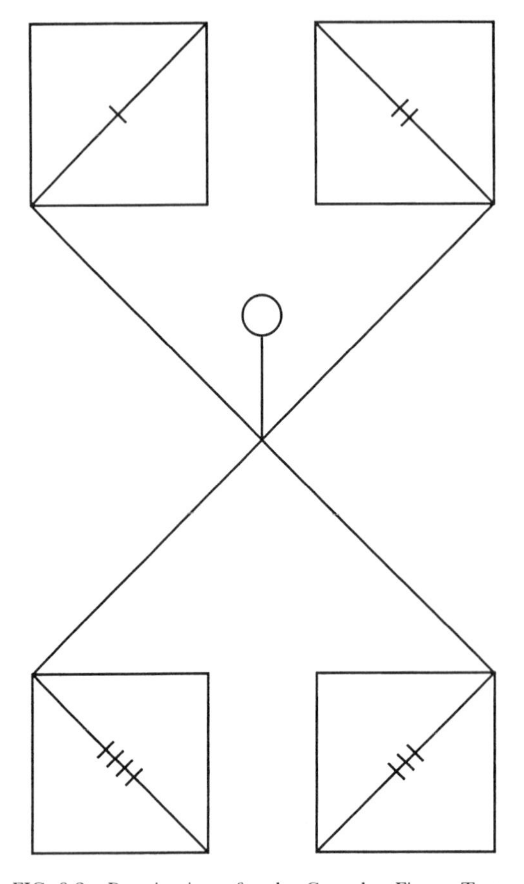

FIG. 9.2. Practice item for the Complex Figure Test.

As a section of the drawing is completed, hand the subject a different pen and take back the one the subject is using. To keep track of the sequence, use the pen you have just taken back to write the figure 1 in a corner of the paper the subject is using. When you change pens again, use the second pen to write 2 underneath the 1 and so on. You will then have a color-coded key indicating the sequence in which the various elements of the drawing were executed.

When the subject has finished the practice item, say:

*Let's have a look now to see if your drawing looks like this one you copied.*

The examiner talks through any discrepancies with the subject, for example:

*You see, this little box over here* [point to the original] *that looks like a house has got three lines in it, but in your drawing, there are only two. Show me how you fix*

> *that. Good. Now can you see any other differences between your drawing and my picture? Show me.*

If necessary, help the subject find other examples. It is not necessary to point out every discrepancy. The purpose is to have the subject understand the importance of exactitude in the copying task. Say:

> *Now that we've practiced, I think you know what to do. Here's the next one I want you to copy* [hand subject the Rey figure]. *Remember you don't have to rush, but you must make your copy look exactly like this one.*

The normal administration of the test proper now follows.

## UNIFORMITY OF TEST ADMINISTRATION

The trade union slogan, "One for all and all for one," is also applicable to psychometric testing, and is readily applied in a research protocol. As shown earlier, a fixed practice protocol is established, and uniformly administered to all subjects regardless of their level of education or prior test experience. Learning is thus maximized in interaction with the test administrator, with less test-wise subjects able to make a greater learning gain, and all subjects in consequence being tested at a point on the learning curve that is closest to their own asymptote. In chapter 10, methods of arriving at similarly standardized procedures for clinical application are described.

Chapter 10

# A Core Test Battery

As noted in chapter 5, clinicians working with clients from culturally different backgrounds feel compelled by reality demands to make immediate use of familiar assessment instruments, even if they have not been validated for use outside the Western cultural enclave. The purpose of this chapter is to discuss the principles of test selection, administration, and interpretation that will allow clinical work to proceed while the task of test validation and norming for clients outside the Western mainstream is in progress.

## A RATIONALE FOR DESIGNATING A CORE TEST BATTERY

Given the huge proliferation of psychological and neuropsychological tests on the one hand (over 500 tests are indexed in the third edition of Lezak's *Neuropsychological Assessment*, 1983), and the limited research resources available for this specialized task, dissipation of effort by unfocused norming ventures will squander resources that could be used to gather scores for core tests for which at least some international norms already exist.

What is needed is a core battery of neuropsychological tests that focus around which researchers can pursue individual interests and at the same time meet the needs of regional and international collaboration by using these core tests as marker tests within their own more extensive batteries. Unfortunately, though "core battery" is a popular term in current neuropsychology, it has no meaning that goes beyond the ad hoc needs and target population of particular research groups. There is a core battery of neuropsychological tests for the assessment of multiple sclerosis (Beatty &

Scott, 1993), a core battery for the Mayo Older Americans Normative Studies (Smith, Ivnick, Malec, & Petersen, 1994), a core battery for the evaluation of clinical trials (Ruff & Crouch, 1991), and several core batteries focus on HIV-related cognitive decline (Butters et al., 1990; Maj et al., 1994). Lezak's basic battery (1995, p. 122) has greater generality but incorporates a number of culture-bound tests, making it unsuitable for use in culturally different settings.

### A Factor Analytic Approach to Designating a Core Test Battery

In the light of the lack of clear criteria for the designation of a core battery, it is tempting to adopt a factorial structure-of-intellect approach in order to define cognitive domains, and give content to these with tests that have high loadings on each of these factors. This approach has been refined through increasingly more sophisticated factor analytic studies (Ekstrom, French, & Harman, 1979; French, 1951; Guilford, 1956), culminating in Carroll's monumental meta-analysis of 450 datasets on which he has based his proposed three-stratum structure of cognitive abilities (1993, chap. 15 and Fig. 15.1). Unfortunately, even at the level of second-stratum factors,[1] Carroll's scheme is not helpful in defining the broad cognitive domains that should be sampled by a test battery (Table 10.1). These problems of test assignation to factors multiply as the second-level factors are decomposed into their component abilities.

A more serious difficulty is that the diagnostic utility of a test battery constructed so as to match tests to an identified factor structure is unclear. For example, a South African equivalent of the Wechsler Preschool and Primary Scale of Intelligence–Revised (WPPSI–R), the Junior South African Intelligence Scale, using Guilford's 1956 Structure-of-Intellect (SOI) model, was commisioned in 1967 and published by the South African Human Sciences Research Council in 1981. Ability tests that matched each cell of the SOI model were devised, resulting in an lengthy battery of 22 subtests. Later factor analysis showed that a subset of 8 subtests was as reliable as the full battery, again calling into question the clinical utility of factor analytic models of ability.

### An Empirical Basis for Designating a Core Test Battery

*The Wechsler tests* (the WPPSI–R, WISC–III, WAIS–III, and WMS–III) continue to be the most widely used in neuropsychological practice, and have generated a large body of quantitative and process-oriented studies with both

---

[1]These are Fluid Intelligence, Crystallized Intelligence, General Memory and Learning, Broad Visual Perception, Broad Auditory Perception, Broad Retrieval Ability, Broad Cognitive Speediness, and Processing Speed (RT Decision Speed).

TABLE 10.1
Carroll's Structure of Cognitive Abilities in Relation to Neuropsychological Test Domains

| *Second Order Cognitive Factors* | *Neuropsychological Test Domains* |
|---|---|
| Fluid intelligence (2F) | Problem-solving and abstraction |
| Crystallized intelligence (2C) | Language |
| General memory and learning (2Y) | Verbal memory and learning<br>Visual memory and learning |
| Broad visual perception (2V) | Visuoconstructional abilities |
| Broad auditory perception (2U) | |
| Broad retrieval ability (2R) | |
| Broad cognitive speediness (2S) | Speed of processing |
| Processing speed (RT Decision speed) (2T) | Speed of processing |
| | Stimulus resistance |
| | Attention and concentration |
| | Executive function |

adults and children. Lezak (1983) noted that "an enormous body of knowledge has grown up around the Wechsler tests"[2] (p. 241), and that greater testing efficiency may not compensate for the "hard-won achievements of familiarity and experience" (1983, p. 242; 1995, p. 689). Test norming is a market-driven enterprise, and because the Wechsler tests have meticulous norms for North America and most Western countries, they provide a useful comparative platform for standardization in the developing countries.

***Marker Tests.*** The authors of the World Health Organization Neurobehavioral Core Test Battery have urged that its seven tests be included "as standard marker tests within larger batteries to allow cross-cultural comparisons between studies and countries" (WHO, 1986, p. 4). As shown in chapter 2, the most extensive developing country database that exists at the present time has been gathered for this brief battery. Six of its component tests have been taken up in the recommendations that follow, and marked with an asterisk in Table 10.2. The Profile of Mood States has been omitted because of the state–trait confusion it caused in our South African studies of semiliterate paint factory workers (see chap. 5, and Nell et al., 1993, p. 310, for a fuller discussion of this problem).

On this basis, 11 of the 14 WAIS–III subtests have been included in the proposed battery, 12 of the 17 WMS–III subtests (five optional subtests have been omitted), and six of the WHO–NCTB tests.

[2]WAIS* appears in the titles of 898 articles in the PsycLIT database, WISC* in 1,214 titles, and WPPSI* in 112.

## INNOVATIONS IN THE WAIS–III AND WMS–III

It is of particular interest to neuropsychologists that the Psychological Corporation engaged an advisory panel of leading North American neuropsychologists chaired by the late Nelson Butters to guide the revision process that led to publication of the WAIS–III and the WMS–III. As a result, both scales now incorporate methods that have for decades been unique to clinical neuropsychology, and thus bring elements of the factorial and diagnostic precision of neuropsychological methods to mainstream psychological assessment. The benefits of this approach are especially marked in working with clients from culturally different backgrounds, and for this reason, the principle subtest and factor structure innovations in these two tests are now reviewed in some detail.

### Extended Practice in the WAIS–III

To the great advantage of psychologists working in the developing countries, the underlying principles of guided learning and extended practice have implicitly been adopted in the WAIS–III, not because the test compilers recognized the need for guided learning, but in order to lower the floor level of the subtests. This has been done by introducing between three and five *reversal items* at the beginning of the Picture Completion, Vocabulary, Similarities, Block Design, Arithmetic, Matrix Reasoning, Information, and Comprehension subtests.

Arithmetic, for example, begins with Item 5; if the client fails either Item 5 or Item 6—these are the two *basal* items—the four reversal items are administered in reverse sequence, continuing until two successive items are answered correctly. To give a feel for the learning gradient, these reversal items for Arithmetic, in sequence from Item 1 to Item 5, are to count three blocks on the table, to count seven blocks, to count the remaining blocks after two are taken away, to subtract 1 from 3 and finally to add 5 and 4.[3]

Because of the inclusion of these easier items, the maximum age has been increased from 74 in the WAIS–R to 89 in the WAIS–III. As a result, the scale has gained greater clinical utility in the diagnosis of mild to moderate mental retardation, and therefore of both traumatic and degenerative brain damage (such as Parkinson's, Alzheimer's, and Huntington's diseases), which fall within the purview of neuropsychologists.

Serendipitously, however, and with no such overt intention on the part of the test compilers, the WAIS–III has become a substantially more cli-

[3]The administration sequence in the test instructions, as the name "reversal items" implies, is from Item 4 back to Item 1, or until two successive questions are correctly answered.

ent-friendly instrument in non-Western settings. By using the reversal items in normal rather than reverse sequence at the beginning of each of the subtests in which they appear, the examiner creates a gentle and non-threatening learning opportunity for non-test-wise clients to understand test demands and prepare gradually for the more difficult later items. And, because full credit is given for the reversal items, whether or not they are administered, the comparability of standard scores is not affected by this change in the administration procedure.

An equally welcome change is the introduction of extended practice on the Block Design subtest (the details of this change are reviewed in "Index Domains" later)—although again this arises because of the psychometric need to lower the test's floor level rather than to cater for culturally different clients.

### From Psychometrics to Information Processing

A further notable development in the WISC–III and the WMS–III—both products of the psychometric tradition—is the incorporation of a factor-analytic information-processing approach. The resulting index scores achieve at least an approximation to the diagnostic precision hitherto attainable only through lengthy test batteries developed within the cognitive processing framework and more appropriate for research than clinical applications.

***Factor Analysis of the WAIS–III.*** Confirmatory factor analysis shows that a four-factor model has the best fit with the the WAIS–III standardization sample and for all but one of the age bands (Psychological Corporation, 1997, p. 110). These factors have been elaborated as four index scores, each with a mean of 100 and a standard deviation of 15, as for the full scale IQ scores. Diagnostic precision has been enhanced by the introduction of three new subtests: Matrix Reasoning, which contributes to the Perceptual Organization Index; Letter–Number Sequencing (in the Working Memory Index); and Symbol Search (in the Processing Speed Index). The indices and the 11 subtests from which they are derived are Verbal Comprehension (to which the Vocabulary, Similarities, and Information Subtests contribute); Perceptual Organization (Picture Completion, Block Design, and Matrix Reasoning); Working Memory (Arithmetic, Digit Span, and Letter–Number Sequencing); and Processing Speed (Digit Symbol—Coding, and Symbol Search). These indices, as with those summarized below for the WMS–III, are reviewed in the *Technical Manual* for the WAIS–III and the WMS–III (Psychological Corporation, 1997).

Note that the Verbal Comprehension Index has been excluded from the core test battery: As noted in chapter 5, tests that carry heavy cultural baggage cannot be "adapted" by translation, on the assumption that all cultural effects reside in language.

***Factor Analysis of the WMS–III.*** Confirmatory factor analysis gives the best fit to a five-factor model. These factors and their component subtests are Auditory Immediate Memory (Logical Memory I and Verbal Paired Associates I); Visual Immediate Memory (Faces I and Family Pictures I); Auditory Delayed Memory (Logical Memory II and Verbal Paired Associates II); Visual Delayed Memory (Faces II and Family Pictures II); and Auditory Recognition Memory (Logical Memory II—Recognition and Verbal Paired Associates II—Recognition).

There is also a supplementary index for delayed auditory recognition, which is thus distinguished from retrieval, and four diagnostically useful auditory process scores relating to the efficiency of learning and retrieval. A general memory index is computed by summing the visual and verbal delayed recall tasks.

The new WMS–III subtests are Faces and Family Pictures (from which the Visual Memory Index is derived); Letter–Number Sequencing (identical to the WAIS–III test and a component of the Working Memory Index); and Word Lists, which is an optional test.

Figural Memory and Visual Paired Associates in the WMS–R are replaced by Faces and Family Pictures (see later), which are less susceptible to verbal encoding, and because "there is little empirical evidence that the WMS–R visual memory subtests are adequate measures of a hypothetical 'pure' visual memory system or that they are differentially sensitive for individuals with unilateral hemispheric lesions" (Psychological Corporation, 1997, p. 16).

The Family Pictures subtest has not been taken up in the core test battery because the activities shown in the color pictures of these White middle-class Americans will be incomprehensible to clients unfamiliar with daily life in the United States. This is regrettable, because the scoring system is simple and attractive, and the test would make a welcome contribution to memory testing if the sketches were redrawn to reflect more widely comprehensible content.

***Co-norming the WMS–III with the WAIS–III.*** In the United States, the WAIS–III and the WMS–III, which are complementary procedures, were co-normed. Despite funding limitations in less privileged settings, there is a strong case for following this route, because judgments about the integrity or impairment of memory (to which the Wechsler intelligence scales are blind) lead to decisions of great moment for the individual. In this area and in the absence of objective criteria for the evaluation of memory, to which clinical intuition and experience may be poor guides, insurance companies may be tempted to pronounce genuinely impaired persons as fit to continue in employment, or as not entitled to personal injury compensation.

## PRINCIPLES OF TEST ADMINISTRATION

As chapter 9 has shown, the psychometrician's problem in dealing with non-test-wise clients is how to bring the zone of proximal development into testing: How, in other words, does one set about moving non-test-wise clients toward a situation in which they can fairly compete with others who have had more formal education or greater exposure to testing?

### Extended Practice

A test is clearly not a test if it has no novelty. The goal of extended practice is not to replace the novelty of the test items by automatized responses, nor is that the intention of Kendall et al. (1988) when they asked whether or not "in any one test exposure, one is measuring [a fully developed ability, or] at a point on the acquisition curve for the ability being measured?" (p. 308). On the contrary, it is to level the playing field, or make it as nearly level as its perhaps intransigent contours will allow, by bringing the non-test-wise client to a point of familiarity with the test apparatus and the task demands that is equivalent to that taken for granted by the test author. Discussing the factor structure of elementary cognitive tasks (ECTs), Carroll (1993) made essentially the same point: "Many ECTs require considerable practice before they 'stabilize' in the sense of producing unchanging intertrial correlations and variances, apart from measurement errors." It is quite posible, he continued, that the factor structure of ECT decision time, for example, "could be attributable to individual differences in the rate and degree to which subjects were able to adapt to the requirements of any given ECT" (p. 506).

In an ideal world, completion of the learning experience embodied in the period of extended practice—it is the executive that is being educated—will bring clients to their asymptotic performance before the test proper begins. The quantitative indicator that this has been achieved is as Carroll defined earlier—that apart from measurement errors, subjects produce unchanging intertrial correlations and variances.

***Practice Procedures.*** It is helpful to think of each extended practice session as divided into *practice* and *speeded practice* phases as set out in chapter 9. In summary, the emphasis during practice is on familiarization with the test materials or apparatus, and the test demands with regard to speed, accuracy, self-correction, and so on (Sternberg called this task acquisition). The test administrator applies the principles of guided learning to this phase by talking the task through with the client and offering as much assistance as possible. In the speeded practice phase, the goal is rapid, error-free performance: The examiner encourages the client to work as rapidly *and accurately* as possible, then withdraws from active coaching.

Extended practice is recommended for all clients who are not fully test-wise—even though they may have completed 12 or more years of school. This will apply, for example, to the victims of a failed educational system in which the teachers are unqualified, classes too large, and pass marks given because the student has become too old for comfort (the South African phrase is, "He was condoned to Standard 5/to high school") or was able to buy the exam paper. But if the test-wise criterion is met, as it will be in persons who have been exposed to a demanding educational system in which achievement-type tests were often used, instructions and practice remain as specified in the test manual. A safe rule is to give additional practice when in doubt.

***WHO–NCTB.*** Should extra practice be given to non-test-wise clients on the six WHO–NCTB tests? It is advisable to do so despite the comparison difficulties this introduces with regard to the existing WHO–NCTB database. The reason is that if the scores of individuals on tests with complicated instructions such as the Benton Visual Retention Test (which in our experience non-test-wise clients often misunderstand), or digits backward (see later) are recorded before these clients have understood what is expected of them, the examiner will be recording scores unrelated to the client's true abilities, so that impairment may be attributed to populations where it does not exist. Rather than continuing to record possibly invalid scores worldwide, it is recommended that a uniform extended practice procedure (as proposed later) be adopted for the WHO–NCTB.

## Unpacking Hidden Meanings

***Self-Correction.*** Test-wise clients know perfectly well that if they give an answer and the examiner says nothing—especially if the examiner won't make eye contact with them!—they had better try again; or, if the examiner looks up from the stopwatch and says, "OK, let's try another one," then it means that the time limit has expired. These subtexts are unknown to non-test-wise clients, and naive examiners can make foolish attributions. For example, suppose the neuropsychologist notes as follows in the report:

> Mr. Bloggs made many errors on the Arithmetic subtest, but made no attempt to use the remaining time in order to self-correct. Instead, he stared aimlessly out of the window or fiddled with apparatus items on the shelf next to him. This raises the possibility that a motivational deficit with poor self-monitoring is present. Mr. Bloggs should not therefore retain his present employment as calibration foreman, but be placed in a less demanding position.

Nonsense! Poor Mr. Bloggs quite sensibly assumed that if the examiner had temporarily lost interest in him, good manners required him to wait patiently until this important person was ready to continue. Unless the hidden meanings are unpacked, Mr. Bloggs will not have the faintest idea that the examiner's thoughtful silence is intended to encourage him to take another stab at the correct answer.

During the first appropriate practice trial, it is therefore essential for the examiner to say something along the following lines:

> Remember that when the test begins, I'm not allowed to help you. If you give the right answer, I'll go on to the next item, but if you get it wrong—or if you don't give an answer at all—I'll wait until the time is up and then go on with the next item in the test. So please remember that if I say nothing, you should think again about your answer. Now let's try that on the next one [this will usually be the first of the speeded practice items]. I'll pretend your answer is wrong, and let's see what you do.

*Time limits* are similarly mysterious to non-test-wise clients. Examiners must therefore be explicit about the kind of timing applied to a test. If the item is untimed (e.g., design copying, Raven's Matrices, and memory tests), the examiner should say so; if there is no time limit but extra points are given for completing a larger number of items, as on Coding or Symbol Search, the examiner should explain this; similarly, clients must be told if there is a time limit but there are also extra points for speediness, as on Block Design or Object Assembly and some Arithmetic items, or if there is just a flat limit, as on Picture Completion. In each of these different systems, the examiner should be explicit about how the time limit is applied. It is not a good idea to go to the next item when the time limit expires without saying anything. Rather, say "Time's up!" in order to emphasize once again the need for swiftness.

### Computer-Administered Tests

The chronometric, computational, and time-efficiency benefits of computerized testing are overwhelming, and this modality is increasingly used in the developed countries. Recognizing the importance of reaction time measures, the WHO–NCTB incorporates a battery-powered timer (although this instrument lacks the capability for choice reaction time measurement). The availability of battery-powered laptop computers means that computerized testing is readily available for clients in non-Western settings.

However, many of these individuals have poor reading skills, and few are familiar with computers, so that self-administered computer testing is not feasible. Moreover, the computer keyboard and screen are likely to be intimidating to semiliterate, rural testees, whose only prior computer-type experience may have been with arcade games now increasingly common in shops in rural settings. One way of addressing this problem, as in the Performance Probe Battery, is by using specially constructed response boxes wired to the keyboard, reserving the keyboard itself for use only by the test administrator. For True–False or Yes–No responses, a two-button box was provided, with the "Yes" button colored green and the "No" red: The explanatory pattern compared these buttons to a traffic signal, green for "yes you can" and red for "no you can't." For tests involving a directional stimulus, an elongated Left–Right box about the same length as the screen was used, with the one button located at each end.

## A CORE NEUROPSYCHOLOGICAL TEST BATTERY

The factorially derived indices for the WAIS–III and the WMS–III are important for two reasons: First, they provide a model for the confirmatory factor analysis that is an essential aspect of construct validation (see the concluding section of this chapter). Second, these indices have a better fit with the neuropsychological assessment process than the factor analytic model reviewed earlier (Table 10.1), and provide a framework around which comprehensive assessments can be structured. Table 10.2 sets out a proposed core test battery for adults with with less than 12 years education, and the hypothesized factor analysis structure of this battery, further discussed at the end of the chapter. Table 10.3 gives some useful supplementary tests for adults and children.

Each of the hypothesized factorial domains in the core test battery set out in Table 10.2 is now briefly reviewed; the next section, "Supplementary Tests," reviews the materials in Table 10.3. Reader familiarity with the neuropsychological test armamentarium is assumed, and tests are described only when this is necessary for understanding of the selection rationale or details of administration and practice. The test descriptions and commentary that follow are thus not intended as a comprehensive test manual—that purpose is served by many excellent assessment handbooks—but as a brief guide to the examiner's special concerns in the assessment of culturally different clients, in particular the accessibility of the proposed tests for such clients and extended practice needs.

### Visuomotor Abilities

A pegboard test is essential for the information it gives about the client's ability to cope with the simple motor tasks required for activities of daily living, and because it is directly analogous to many workplace activities for

TABLE 10.2
A Core Test Battery for Adults with Less Than 12 Years Education and Hypothesized Factor Structure

Visuomotor Abilities
- Grooved Pegboard Test
- Santa Ana Pegboard*[a]
- Pursuit Aiming*

•[b] Processing Speed
- Simple reaction time*
- +[c] Digit Symbol (WAIS–III)*
- +Symbol Search (WAIS–III)

• Perceptual Organization
- +Picture Completion (WAIS–III)
- +Block Design (WAIS–III)
- +Matrix Reasoning (WAIS–III)
- Object Assembly (WAIS–III)
- Visual Reproduction I (WMS–III)

• Working Memory
- +Digit Span (WAIS–III and WMS–III[d])*
- +Arithmetic (WAIS–III)
- +Letter–Number Sequencing (WAIS–III and WMS–III[d])
- +Spatial Span (WMS–III)

• Auditory Memory (Immediate)
- +Verbal Paired Associates I (WMS–III)
- +Logical Memory I (WMS–III)

• Auditory Memory (Delayed)
- +Verbal Paired Associates II (WMS–III)
- +Logical Memory II (WMS–III)

• Auditory Recognition Memory
- +Logical Memory II (WMS–III) (Recognition)
- +Verbal Paired Associates II (WMS–III) (Recognition)

• Visual Memory (Immediate and Delayed)
- +Faces I (WMS–III)
- +Faces II (WMS–III)
- Benton Visual Retention Test*
- Visual Reproduction II (WMS–III)

[a]Included in the World Health Organization Neuropsychological Core Test Battery (WHO–NCTB).
[b]Denotes WAIS–III or WMS–III index scores derived from confirmatory factor analysis.
[c]Subtests identified by confirmatory factor analysis as contributing to the Index Score.
[d]The tests are identical.

TABLE 10.3
Useful Supplementary Tests for Adults with Less Than 12 Years Education and for Children, with Hypothesized Factor Structure

| | |
|---|---|
| Visuomotor Abilities | |
| Animal Pegs (WPPSI–R) | |
| Visuopraxis | |
| Geometric Design Reproduction: | One-dimensional |
| | Three-dimensional |
| | Kiddie |
| Complex Figure Test | |
| Clock Faces: Tell the time | |
| Stimulus Resistance | |
| Echopraxis | |
| Trail Making | |
| Mazes (WPPSI–R, WISC–III) | |
| Tower of London/New Tower of London | |
| Austin Maze | |
| Working Memory | |
| Serial Threes/Serial Sevens | |
| Auditory Memory (Immediate, Delayed, and Recognition Memory) | |
| Four Words in Sequence Test | |
| Fuld Object Memory Test | |
| Auditory Verbal Learning Test (Immediate verbal memory: Trials 1–7) | |
| Auditory Verbal Learning Test (Delayed verbal memory: Trial 8) | |
| Auditory Verbal Learning Test (Recognition Memory) | |
| Visual memory (Immediate and Delayed) | |
| Geometric Design Reproduction: Immediate and Delayed Recall | |
| Complex Figure Test: Immediate and Delayed Recall | |
| Language | |
| Oral Word Generation: By category (Animals/Fruit and Vegetables/Articles of Clothing) | |
| Writing to dictation: | |
| Telephone numbers | |
| Words and sentences | |
| Token Test | |

both clerical and manual workers. The two strongest candidates for inclusion are the *Grooved Pegboard Test* (GPT) and the *Santa Ana Pegboard,* originated by Fleishman (1954) and modified at the Institute of Occupational Health in Helsinki. The first enjoys near universal use in neuropsychological batteries and has an extensive existing norm base, and for the second a wide range of scores is available because of its inclusion in the WHO–NCTB. In terms of convenience, these tests are at opposite poles. The GPT fits into a coat pocket, whereas the Santa Ana is twice body width and correspondingly heavy.

The advantages of the Grooved Pegboard are that the flanged pegs, about the size of a half-inch nail, must be manipulated in order to fit them into grooved slots, testing fine motor dexterity, detecting resting or intention tremor, and also showing sensitivity to impaired stereognosis or proprioception. By contrast, the Santa Ana consists of a base plate with 48 square depressions and 48 cylindrical pegs with a square base. The pegs are large, about an inch in diameter, and all require the same 180-degree turn to fit the square peg base to the board; the preferred and nonpreferred hands are timed separately (see Appendix 5). The task is analogous to many factory operations, which is probably the reason it was selected, but does not provide the fine skills information yield of the GPT. It is therefore suggested that for clinical testing, preference be given to the GPT, but that for research applications, both the Santa Ana and the GPT be used in order to contribute to the WHO database.

For *extended practice* on both the GPT and the Santa Ana, allow clients to insert 10 pegs for practice, followed by another 5 for speeded practice. On both tests, discontinue if you see that the client cannot do the task (e.g., because of a tremor or paresis).

The Fleishman (1954) Pursuit Aiming Test[4] is a dotting task that relates to an aiming factor defined as the precise execution of "a series of movements requiring eye-hand coordination [or] the correct positioning of a pencil mark on paper" (French, 1951, cited in Carroll, 1993, p. 536). For *extended practice,* make an additional copy of the test proper, using the first two rows (60 dots) for practice, and the next row (30 dots) for speeded practice (Appendix 6). The practice material given on the front of the test blank should not be used—this differs in format from the test proper, and there are not enough items for adequate practice.

## Processing Speed

***Simple Reaction Time.*** Carroll observed that time limited power[5] tests "are seriously biased against individuals with low rates of test performance" (1993, p. 508). This bias would by definition extend to clients for whom deliberation is a more salient value than speed (see Sternberg's [1984,

---

[4]This and other factorially precise tasks operationalized in the French Reference Kits (Carroll, 1993, Table 3.1; Ekstrom et al., 1979) have dropped out of the neuropsychological testing repertoire, and it is to the credit of the developers of the WHO–NCTB that they have reintroduced some of these measures.

[5]"A *pure speed* test is one in which individual differences depend entirely on speed of performance. Such a test is constructed from items of uniformly low difficulty, all of which are well within the ability level of the persons for whom the test is designed. The time limit is made so short that no-one can finish all the items. A *pure power* test on the other hand has a time limit long enough to permit everyone to attempt all items. . . . The test includes some items too difficult for anyone to solve, so no one can get a perfect score" (Anastasi, 1988, cited in Carroll, 1993, p. 444).

1988] characterization of the experiential archon), and most of those who are not test-wise. What is one then to make of Thorndike's aphorism (cited in Carroll, 1983) that "other things being equal, the more quickly a person produces the correct response, the greater is his [her] intelligence"? Carroll correctly dismissed this as a societal judgment no longer supported by the data: The correlations between speed and power are too small and too variable "to justify any hope that rates of performance on elementary cognitive tasks could be used as indicants of intelligence level" (p. 508). Matters are further complicated by the observation that reaction times cannot be taken as a univariate measure of apprehension or pure speediness, because the data reviewed by Carroll show that the more complex the task on which reaction time is measured, the longer the reaction time and the greater the correlation between reaction times and level (i.e., power) abilities.[6]

Notwithstanding these caveats, reaction time remains the most fundamental and one of the most useful of the elementary cognitive tasks (Brody, 1992, chap. 3; Carroll, 1983, chap. 11). Simple reaction time (SRT) to visual stimuli has become a standard component of neurotoxicology investigations, and has thus been incorpoated in the core battery. For the measurement of SRT, the WHO–NCTB *Operational Manual* recommends use of the portable battery/mains-powered Terry Reaction Time Device (Appendix 7), which stores and displays reaction times, range, number of responses, and standard deviation. Because it is difficult to compare reaction time data from different test devices, it is recommended that this instrument be used in the interests of international comparability. The greater sophistication of the Hick Box, which allows for refined measures of simple and choice reaction times to be computed by subtracting movement time from the overall response time (Jensen, 1988), is offset by the cumbersome apparatus required and the uncertain comparability of Hick Box reaction times with those obtained by the Terry instrument and similar devices.

Even on this most basic of performance processes that is of interest to psychologists, cultural influences are strong: As Table 2.1 shows, reaction times in South America and South Africa are substantially slower than in Europe. In this light, *extended practice* is essential. The recommended procedure is that adopted in the Performance Probe Battery: The importance of speed is emphasized by framing the task as a conflict. Depending on the cultural background of the client, the metaphor might be a street fight, a confrontation with a poisonous snake, or a scene from a Wild West cowboy movie (as set out in chap. 9).

---

[6]This is in fact a restatement of Hick's Law (1952): Latency of response to a given stimulus is linearly related to the number of binary digits or bits of information (by which information content or uncertainty may be quantitatively defined) that are presented to the client.

***Digit Symbol Substitution.*** The diagnostic and predictive utility of this subtest (its equivalent in the WISC–III is Coding) are well established, again arising from this procedure's obvious parallels with everyday workplace tasks—visual scanning, fine motor control, and incidental learning. It is well-suited for use with non-test-wise clients, even those who are semiliterate, because the task is not to write numbers, which is an entrenched skill for literate people, but to write novel symbols, which is nonentrenched for both literates and semiliterates. On the other hand, the Symbol Digit Modalities Test is unsuitable for semiliterates, because their less-developed skills in writing figures place them at a relative disadvantage on what is ostensibly a test of motor speed. For *extended practice,* make an additional copy of the test blank, using the the last three rows: two for practice and the third for speeded practice.

*Target detection tasks* have long been used in neuropsychological assessment, among them cancellation of designated letters, digits, or symbols. Because of their limited experience with letters and numbers, symbol detection has the relatively low task familiarization demands for non-test-wise clients. *Symbol Search* first appeared in the WISC–III as a 45-item test. The WAIS–III version has 60 items with a 2-minute time limit; this is too brief to qualify as a vigilance task, but nonetheless a useful measure of concentration and information-processing speed. Evidence of how thoroughly the floor level of the WAIS–III has been adjusted is that the typeface is larger than in the WISC–III, and there are now more sample and practice items than in the children's test. Because it is so easy to jury-rig one's own cancellation or targeting test, standardization in this domain is poor. It is thus recommended that standardization focus on the WAIS–III version of Symbol Search.

## Perceptual Organization

Picture Completion, like Block Design and Object Assembly, is an old staple of intelligence testing, dating back to the 1917 Pintner and Paterson test[7] (Frank, 1983). The reason for this popularity is that all three of these procedures access basic adaptive skills—pattern matching, making a whole out of seemingly unrelated parts. For Picture Completion, the task is to scan a complex array and determine which necessary part of the whole is missing; the inverse of this procedure is a test known as Absurdities. This derives from the Stanford–Binet and has been especially popular with South African test

[7]All 11 of the subtests taken up by David Wechsler in the 1939 Wechsler–Bellevue Adult Intelligence Scale—predecessor of the WAIS and its revisions—had already been applied on a vast scale between 1917 and 1919 in the Army Alpha and Beta tests, of which the Pintner and Paterson had been a forerunner.

developers. Here, the part is not missing but dysfunctional. Thus, a teapot in Picture Completion might be shown witht spout disconnected from the pot; in Absurdities, the spout would be upside down.

The missing part principle is not immediately obvious to non-test-wise clients. A common error is to name a missing part that is external to the given frame. For example, shown a picture of a hand with a missing fingernail, the client might say, "The arm is missing"; for glasses without a bridge, "The person's face is not there"; or for a picture of a ladder with a rung missing, "There is no one to climb the ladder."

*Practice* is thus essential in order to familiarize both adult and child clients with the requirement that the missing part must be a significant element of the whole, and that it must be missing from the picture as given. Four practice items are shown in Fig. 10.1, all of which are susceptible to an interpretation that goes beyond the picture itself—clients may for

FIG. 10.1. Four practice items for Picture Completion.

example say that the horse's body is missing, that the bicycle has no rider, etc.. Clients are talked through the first two practice items, the one-eared horse and the teapot, bearing in mind that even if the above difficulties do not at once emerge, they lie latent; on the the next two speeded practice items (bicycle and wristwatch), the examiner emphasizes that there is a time limit and the answer must be found rapidly.

Both Block Design and Object Assembly present novel problem-solving tasks with manifold daily life implications. Although correctly regarded as measures of $g_f$, fluid intelligence, it is wrong to assume that because these are novel tasks to all clients, they make similar demands of both literate and semiliterate clients. Schooling and a stimulus-rich home environment allow literate clients to become familiar with spatial construction tasks through games such as Lego, Meccano (a staple of 1940s and 1950s childhoods), Tinkertoy, and jigsaw puzzles. Although no clients will have done the specific Block Design and Object Assembly tasks in the tests (unless they are being retested), the novelty of these tasks is thus lower for literates and older children, but higher for semiliterates and younger children. Extended practice is therefore imperative.

***Block Design.*** As already noted, the WAIS–III administration procedure for this subtest has been radically modified to provide a graduated learning experience. There are two trials for each of the four reversal and two basal items, giving non-test-wise clients 12 learning opportunities: If the first trial on any of these seven items is failed, the examiner says, "Watch me again," rebuilds the model, scrambles the blocks, and says, "Now you try it again and be sure to make it just like mine." Although this method gives extended practice, it does not constitute guided learning, in which the examiner patiently coaches the client through errors, encouraging self-correction at each step, as illustrated by the examples in chapter 9.

However, these first five WAIS–III designs fail to take into account a feature of the block design task on which many investigators have remarked, namely, that there are two classes of design. Walsh (1985, p. 15) used the terms *nonembedded* and *embedded*. Nonembedded relates to designs that can be built up by matching each block to a separate and readily identifiable quadrant or one ninth of the design. On the WAIS–III, these are the first five items that are built from block models, which of course show the divisions between blocks and are therefore by definition nonembedded. The only other nonembedded item is 10.

In the embedded designs, on the other hand, the color flows across block divisions; in other words, these designs do not have a perceptual border that matches any single block. The only practice non-test-wise clients will get on this type of design is thus on the two trials of Item 6. It is therefore necessary to include an additional embedded practice item, which is illustrated in Fig. 10.2. In order to meet the extended practice needs of non-test-wise clients, the recommended procedure is thus to administer Items 1 to 4 of the WAIS–III and also the additional embedded practice item illustrated in Fig. 10.2 as Item 4a. Items 5 and 6 of the WAIS–III are then given as speeded practice.

If your client has difficulty on Items 5 and 6, use a block model to cue block by block, adapting the following interactive method:

> *OK,* (indicate one block on your model) *let's try to get this one right. Show me which block you want to use. Now, let's do this one* (indicate another block on your model). *Where does it go? Let's put these two* (indicate the selected blocks on your model) *together so that your blocks look like mine.*

***Object Assembly.*** Unlike Block Design, in which the embedded designs present special problems, Object Assembly increases in difficulty from item to item in a straightforward linear fashion. The purpose of the additional *practice* item (e.g., the unscored practice item, an apple, from the now obsolete WISC–R) is thus to begin bridging the gap between those familiar with jigsaw puzzle tasks and those for whom this is a novel activity.

Object Assembly is factorially more complex than its cognate tests, does not contribute to the Perceptual Organization Index, and is therefore an optional test. However, it is readily understood by non-test-wise clients, provides an excellent information yield, and has for this reason been retained in the core test battery.

*Matrix Reasoning,* analogous to Raven's Matrices, is a powerful addition to the WAIS–III. It is constructed to represent "four types of nonver-

FIG. 10.2. Additional practice item for Block Design administered as Item 4a.

bal[8] reasoning" (Wechsler, 1997): pattern completion, classification, analogy, and serial reasoning. The subtest is well-suited to the instructional needs of non-test-wise clients. On the three sample items, there is implicit guided learning, because if the client's answer is wrong, the examiner is allowed to explain why. Although additional coaching on the three reversal items is not allowed, these provide further practice and should for non-test-wise clients be given in forward sequence. At its upper level, the test is taxing, with the last three items measuring at above the 97th percentile.

***Visual Reproduction I and II.*** If this test is included in your battery, it is essential that the Rey Complex Figure and its practice item (these are among the supplementary tests in Table 10.3) are administered first in order to provide the necessary practice in the careful and accurate copying of designs, with which non-test-wise clients are unfamiliar—although the WMS–III takes this skill for granted.

### Working Memory

***Arithmetic.*** Arithmetic is not a skill that is restricted to persons with a high school education, but is also ubiquitous among little-educated people who do a great deal of informal arithmetic in their occupations as hawkers, taxi drivers, and the like. The WAIS–III Arithmetic subtest is therefore included in the core test battery, and as noted earlier, the four reversal items provide an excellent extended practice gradient for non-test-wise clients. For children between age 3 and 17, the WPPSI–R and the WISC–III should of course be administered, making full use of the reversal items.

*Digit Span* is a hardy classic of the psychological test repertoire. *Digits forward* (which tests attention, not memory—although many psychologists who ought to know better regard it as a memory test) is readily grasped and executed by semiliterate clients. *Digits backward,* like Letter–Number Sequencing, is an elegant test of working memory in the precise sense of its definition—holding a set of information in memory while a further transformation is applied to it. This is clearly a task that is cognitively distinct from repeating digits forward, and as Lezak repeatedly emphasized, scores for these two cognitively distinct parts of the test should be reported separately.

Our experience shows that there are many advantages to the WAIS–III administration procedure, although regrettably, the WHO–NCTB manual still follows the WAIS instructions; the benefits of the WAIS–III procedure

---

[8]Many studies have shown that these ostensibly visual tasks in fact rely heavily on verbal encoding. Reflecting on one's own thought processes while engaged in working through a Raven's-type problem at once reveals that one is in fact talking one's way through it.

outweigh the comparability drawbacks, and should be followed. The difference is that in the WAIS–III, each span length is repeated twice even if the client gets it correct at the first attempt, and a score between 0 and 2 is given to each trial. The benefits are first that the score is a more sensitive indicator of performance than span length (Nell et al., 1993). Second, for digits forward, this form of administration provides its own practice; there is thus no extended practice for digits forward.

For digits backward, however, *extended practice* is essential. The task should be explained with particular care, first asking the client to reverse the digits 1, 2, then 1, 2, 3. If difficulties emerge at this level, as Lezak suggested (1995, p. 367), it is helpful to point at an imaginary 1-2-3 series in the air in front of you, and then, pointing at the imaginary 3 on the right, say:

> *Now you say 3* (point to the middle position) *-2* (pointing to the left-hand position) *-1.*

Alternatively, write the numbers down so that you can point to each number in turn in the reverse sequence while encouraging the client to say the numbers aloud. Thereafter, practice continues with other 2- or 3-digit sequences until the client has grasped the reversal principle and the test proper can be administered.

I have, however, encountered a number of adult and child clients who do well on the forward task and many other tests, indicating reasonably intact cognitive function, but are quite unable to follow the instruction to reverse the sequence no matter how much time is spent on explanation. The reason for this incapacity is unclear, but once it is established that the client cannot reverse digits, the backward phase should be aborted; the forward score, however, should be reported.

*Letter–Number Sequencing,* an ingenious new task in the WAIS–III, gives a welcome boost to the measurement of working memory. Mixed number–letter strings of increasing length are read to the client, who must rearrange these in sequence, placing numbers first and then letters. For example, the client must transform the string N-5-F-8-R-3 to 3-5-8-F-N-R.

*Spatial Span* in the WMS–III is the visual analogue of digit span (Knox Cubes is an older version of this task). If it is administered after Digit Span, the task demands will be clear to clients, and further practice should not be necessary.

### Auditory Memory (Immediate and Delayed)

***Verbal Paired Associates.*** The easy word pairs, of dubious diagnostic utility in earlier versions of the WMS, have been dropped. All eight pairs are now difficult associates, but concrete and readily imaged. The test is administered and scored as in the WMS–III manual, with the addition that the examiner gives one or two examples, thus:

> *These words come in pairs, like* floor–ceiling. *So if I say* floor–ceiling, *then later I say* floor, *what will you say? OK, let's try a more difficult one this time,* table–blanket. *Then later if I say* table, *what will you say?*

Examiners familiar with the old Wechsler Memory Scale must note that the WMS–III makes provision for selective reminding; for all errors and omissions, the examiner gives the correct word after 5 seconds.

***Logical Memory.*** Robert Miller, the negligent truck driver who skidded off the road in Story B of the WMS–R Logical Memory subtest, has been replaced by a relaxed San Franciscan movie buff. This change is clinically helpful to neuropsychologists working with head injured clients, for whom the Robert Miller story triggered unfortunate memories of their own accident, and in such cases disrupted encoding. More important, Story B is now administered twice in order to ensure that recall is measured on the basis of an adequate opportunity for encoding. In addition, a diagnostically useful measure of thematic recall has been added to the scoring manual.

With regard to the content of memory tests, neuropsychologists working outside North America must avoid confounding narrative memory with the acquisition of what for the client is likely to be meaningless information. Thus, a ship striking a mine (as in the original WMS—to South Africans, a mine is what you find gold in) or a trucker using citizen band radio (in the WMS–R) are great mysteries in other countries,

*Word Lists* is closely modeled on the familiar 15-word Rey Auditory Verbal Learning Test. In the WMS–III version, 12 words are presented over 4 trials, followed by an interference trial, then immediate and delayed recall.

### Auditory Recognition Memory

Recognition is now distinguished from retrieval by including recognition trials for Logical Memory and Verbal Paired Associates (which make up the auditory recognition index), and for Word Lists.

### Visual Memory (Immediate and Delayed)

*Faces* (analogous to the venerable Benton Facial Recognition Test) consists of 24 color photographs of faces exposed for 2 seconds each. There are two recognition trials, one immediately and the next after about 30 minutes, each requiring the client to identify the 24 target faces among 48 photographs.

***Benton Visual Retention Test (BVRT).*** Apart from its utility as one of the marker tests in WHO–NCTB, this is a useful item in its own right. Because it is a recognition procedure, it does not confound memory with drawing ability. We have found that despite its apparent simplicity, the task has high novelty for non-test-wise clients, and that additional practice is essential (Appendix 8). The two practice items are shown in Fig. 10.3. A frequently encountered problem is that clients underestimate the test's complexity, glance at the target item for a second or two, and then withdraw their attention. If this happens, the examiner should say:

> *Look at the patterns all the time I give you. These may be simple patterns, but the others will be more difficult, so be sure to use all the time I give you to study the drawing carefully.*

## SUPPLEMENTARY TESTS

The supplementary tests in Table 10.3 are now reviewed within their hypothesized factorial domains.

### Visuomotor Abilities

***Animal Pegs (WPPSI–R).*** The Grooved Pegboard can be used for children from age 6, but for children up to age 8 from non-Western backgrounds, or even age 10 for children who have less than the age-appropriate schooling, the Animal Pegs subtest of the WPPSI–R is an excellent although multidetermined alternative, because it requires color-to-category matching in addition to its motor component. Clinical experience with African children in Johannesburg indicates that this test is readily understood and well-performed—often at the level of the U.S. norms—and is likely to have good validity in non-Western settings. It has therefore been included in the core battery. In our experience, the standard test instructions provide adequate practice.

### Visuopraxis

These tests entered neuropsychology because of their sensitivity to right hemisphere parietal lesions that give rise to the classic apraxias. Such discrete lesions are seldom seen after traumatic brain injuries, but there are two reasons for including these tests in the battery. In the first place, an apraxia may be present in combination with other deficits, but contribute a substantial additional weight of disability. This is because clients, caregivers, and employers typically remain unable to define or articulate

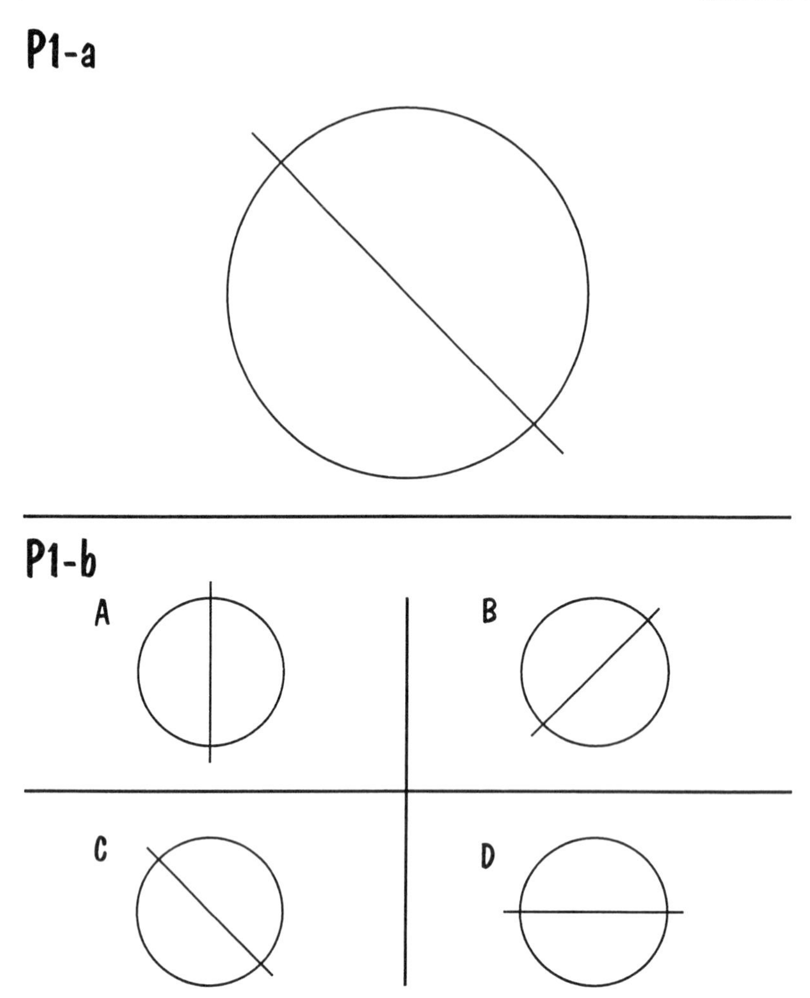


FIG. 10.3. Two practice items for the Benton Visual Retention Test.

these subtle impairments, but are troubled by their inability to perform simple tasks such as understanding a house plan, following a wiring diagram, or drawing a sketch map. Information giving by the neuropsychologist about such impairments can relieve anxiety and initiate sensible remediation.

Second, with both *Geometric Design Reproduction* in its simple and three-dimensional forms, and the *Complex Figure Test* (CFT), the "don't care" attitude that emerges as a result of poor self-monitoring is often revealed by copying tasks, as with the maintenance fitter described in chapter 7 who said his copies of the GDR items were "not 1,000%" but showed little interest in doing better. On both these tests, an immediate recall trial is

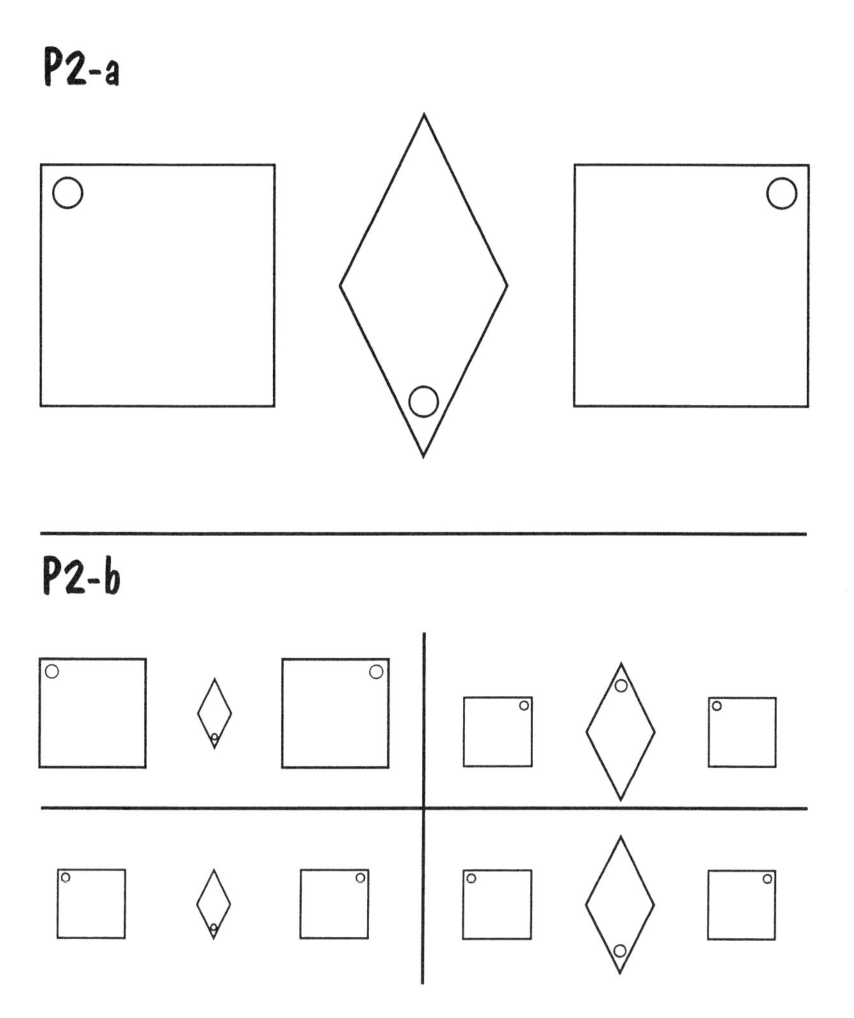


FIG. 10.3. *(continued)*

given directly after the copy phase, and a delayed recall is requested at 30 minutes. These immediate and delayed performances serve to examine *visual memory*. As already noted and in chapter 7, motivational and self-monitoring difficulties can readily be explored by asking clients to comment on the quality of their GDR and CFT copies. In addition, both the immediate and delayed recalls should be qualitatively examined for evidence of such problems.

Two *practice* items, a square and two linked squares, and the five Geometric Design Reproduction items, are shown in Fig. 10.4. In talking clients through the two practice items (Appendix 9), the examiner lays stress on the need for accuracy of shape and size, that the pencil should not be lifted once the item has been started, and that clients can take their time because the test is untimed.

An interesting qualitative aspect of this test, in addition to those previously noted, is the client's planning and use of space when making the copies. For this reason, the designs are not given as a column on the left with neat demarcated spaces on the right for the copies. Instead, they are grouped, leaving their arrangement on the page to the client; this principle

**Two Practice Items for Geometric Design Reproduction**

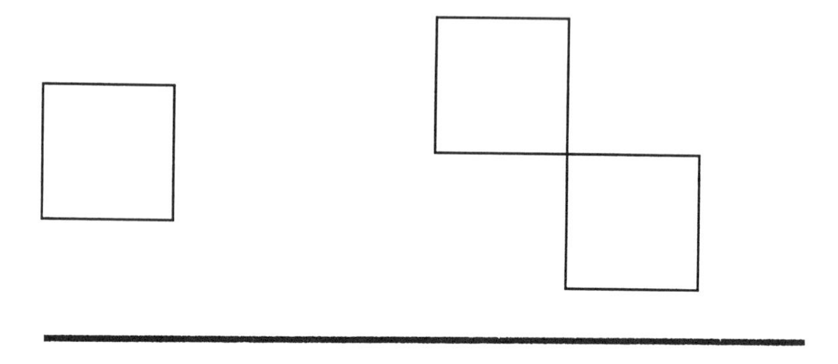

**Five Test Items for Geometric Design Reproduction**

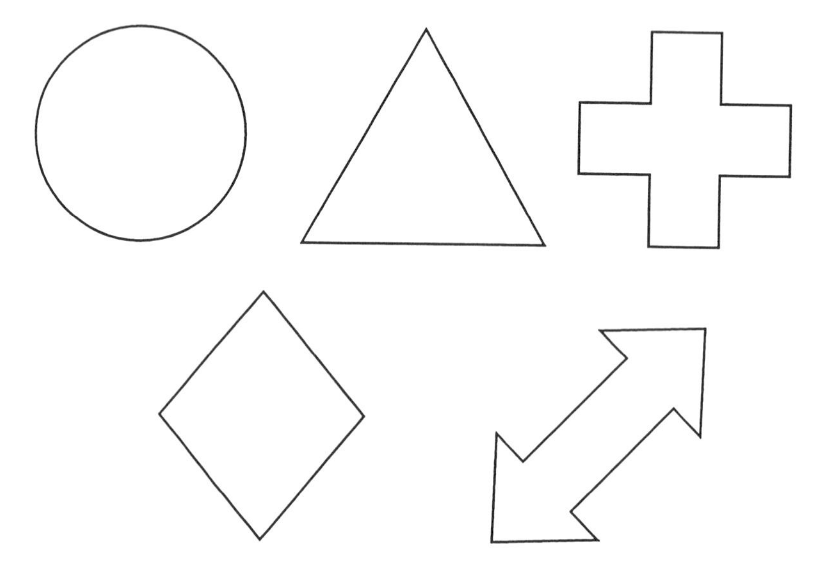

FIG. 10.4. Geometric Design Reproduction (GDR) with two practice items.

is also followed for three-dimensional design reproduction (discussed later).

For the *Complex Figure Test,* Lezak's administration instructions should be followed, including the use of different colored pens in order to monitor the client's planfulness. The administration instructions and *practice* procedure are given in chapter 9, and the practice item illustrated there (Fig. 9.2).

***Three-Dimensional Design Reproduction.*** This set of four designs (Fig. 10.5) is in use at the Karolinska Hospital in Stockholm and has the same diagnostic utility as the two previous copying tasks. However, the difficulty level is higher, making this the copying task of choice for individuals in technical or craft occupations, or planning to train for such careers. Because this version of geometric design is reserved for individuals with some copying skills, practice is restricted to the first item, which is two dimensional and thus at a lesser difficulty level.

*Tell the Time* can with caution be used for semiliterate clients because of the ubiquity of time-keeping and time measurement in urban settings. Ten settings are given in Fig. 10.6, with numbered hour divisions for the first five, and unnumbered hour divisions for the others. The settings get trickier in a more or less linear fashion. In order to avoid confounding

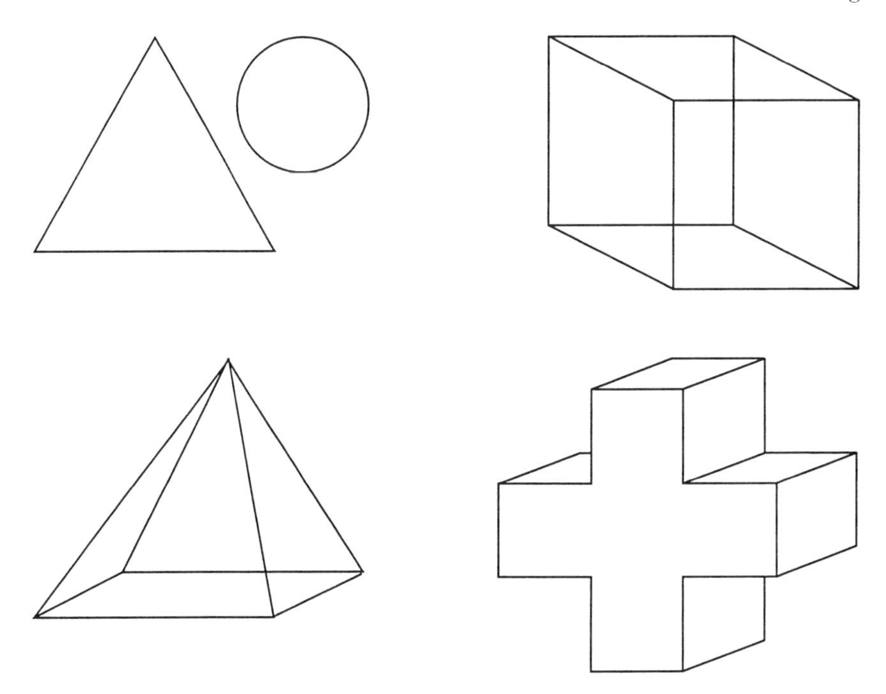

FIG. 10.5. Three-dimensional design reproduction.

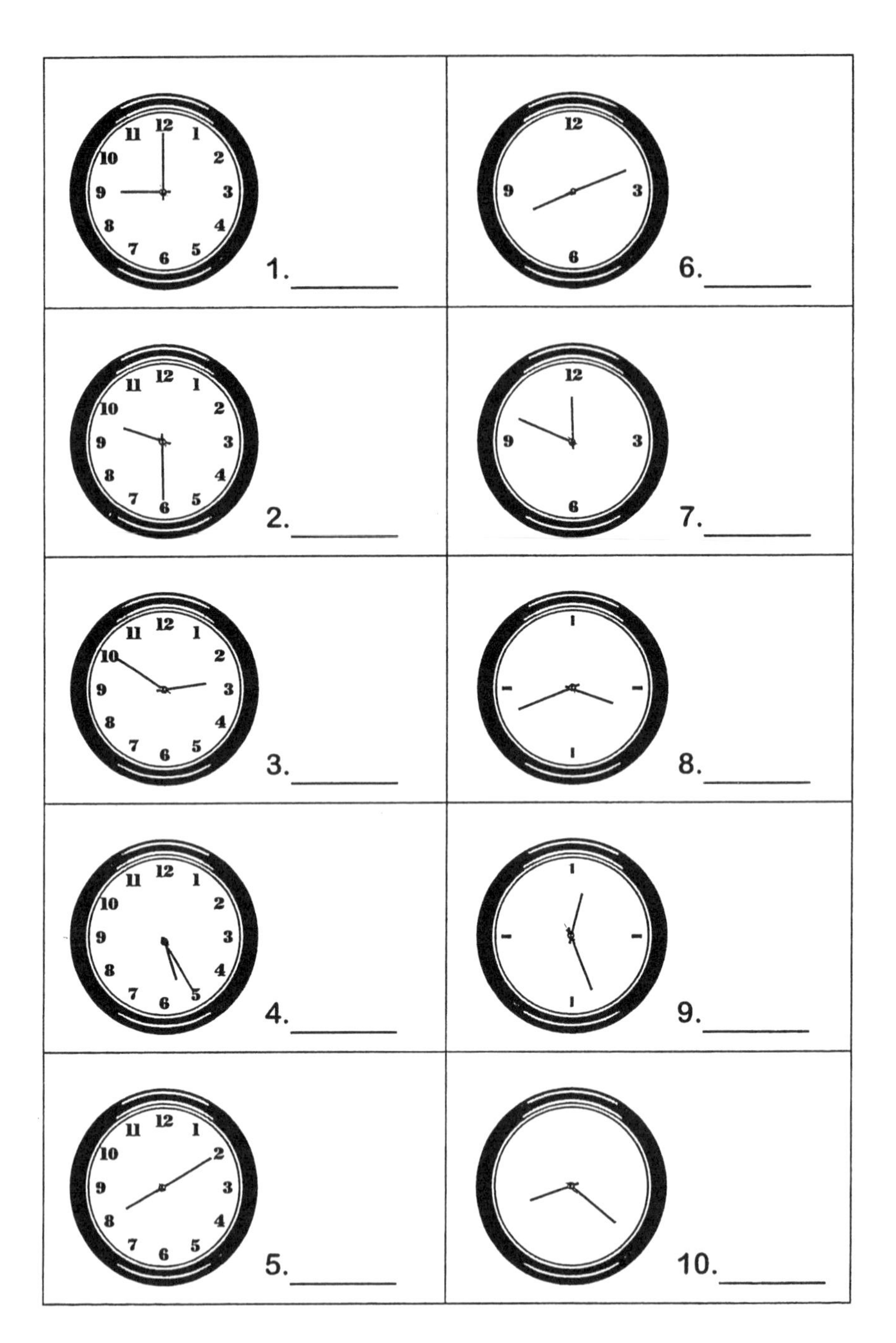


FIG. 10.6. Ten clock faces for Tell the Time.

writing skills with telling the time, the client's answers must always be noted by the examiner.

### Working Memory

*Serial Sevens* (subtracting 7 from 100, again from 93, and so on), and its easier form, *Serial Threes,* are usually given only to clients with 12 or more years education, but with caution can be used with semiliterates who do a great deal of informal arithmetic in their occupations. With clients who have less than 12 years education, begin with Serial Threes, explaining what has to be done and coaching the client over five trials or until the task is being performed fluently; if it seems too easy, you can switch to Serial Sevens, again giving five practice trials or more, if necessary. For *practice,* begin with 99 instead of 100 to avoid practice carryover to the test proper.

### Auditory Memory

***Do Memory Tests Need Practice?*** Four tests in this group are list learning tasks, presented in an approximate difficulty sequence. They provide their own practice, with interpretation focusing on the slope and regularity of the learning curve. On the other hand, there is no practice for the narrative memory task in Paragraph Memory (illogically, Wechsler's term is "Logical Memory"). We have found that the performance of both test-wise and non-test-wise clients is often better on the second of the two stories, suggesting that clients used the first item for task familiarization, enabling them to do better on the second.

***Four Words in Sequence Test.*** There is an excellent information yield from this seemingly trivial test. Four of the words are concrete and readily imaged, and the fourth is abstract—a method used by George Prigatano: The last word is more difficult to encode and typically more often forgotten than the other three. We use *House, Cloud, Tree,* and *Flying* as the four words (an alternative form is *Road, Wind, Door,* and *Calm).* It is helpful to use the record form in Appendix 10, on which the number of trials to acquisition in the correct sequence is recorded, as well as recall at 5, 10, and 30 minutes and at the end of the session. Loss of two or more words at the 5-minute recall suggests that a memory problem of the axial or frontal type may be present.

***Fuld Object Memory Test.*** This is an attractive and nonthreatening test for children and severely impaired adults, who often smile in pleasurable anticipation as they thrust their hand into the bag to retrieve the next mysterious object. The combination of tactile identification and verbal naming consolidate the memory trace, so that poor acquisition is especially diagnostic. Fuld's (1977) procedure calls for both selective reminding and trial-to-trial interference: If the interference is omitted, then the published norms are of course inapplicable.

### Visual Memory (Immediate and Delayed)

An immediate recall trial is administered after Geometric Design Reproduction, Three-Dimensional Design Reproduction, and the Complex Figure Test, followed by a 30-minute test of delayed recall. No indication is given to the client that a delayed recall will be required.

### Stimulus Resistance

The term *stimulus resistance* is used to describe tests in which an absent verbal instruction (absent because it is given once at the beginning of the test) is pitted against a stimulus that is continuously present and therefore more powerful. The exemplar of this capacity in its most elementary form is Luria's motor test of echopraxis (discussed later). More widely, stimulus resistance describes that broad class of tests that includes Trails B, the Stroop Color–Word Interference Task (which is suitable only for fully literate clients), and the planfulness required for successful performance of pencil-and-paper tests, for example, mazes and complex figure drawing; rearrangement tests on which an initial error precludes success, such as the Shallice, 1982, Tower of London test (a simplified variant of the Tower of Hanoi); and tests of orderly, impulse-free learning like the Austin Maze. Walsh coined the term "prehearsal," which nicely captures the cognitive demand made by these tests: The client must resist the "pull" of the stimulus, which is present, and instead follow a more complex—and remote—previous instruction, which is no longer present. This ability rests on preservation of the frontal function that will, in Luria's words, subordinate one's actions to a program by inhibiting "more elementary forms of actions arising in response to direct impressions or as a result of perseveration of previous actions" (Luria, 1962/1980, p. 309).

"Stimulus resistance" is not a recognized cognitive processing factor. Carroll's closest approximation to this construct is in relation to an interpretation of the Stroop effect, namely, the difficulty "of suppressing the highly practised response to the printed word" (1993, p. 491), when clients are for example asked to say "red" when they see the word "green" printed in red ink. Carroll assigned this effect to the the cognitive speed domain, and on these grounds I would hypothesize that this group of tests would load on the Processing Speed factor.

*Echopraxis* (or more correctly the ability to resist echopraxia) is one of the many simple motor tests Luria described in *Higher Cortical Functions in*

*Man* (1962/1980, part 2, chap. 5, "Disturbances that Follow Lesions of the Frontal Regions of the Brain"). The literal meaning of echopraxia is to mimic an action, just as echolalia is the mimicking of sounds. The test instructions are given in Appendix 11.

The *Trail Making Test* entered neuropsychology through the Halstead–Reitan battery. A significant difference between standard scores on Part A and Part B has long been regarded as a good indicator of slowed information processing. Of course, there is no point in administering Part A to a client who cannot do Part B, because diagnostic conclusions are largely drawn by comparing performance on Parts A and B. However, as noted in chapter 4, Trails cannot be administered to clients with less than 12 years education even if they have good language skills and can say the alphabet: Unless there has been sufficient schooling to entrench both the number and letter series, the test will measure an unknown construct.

Both parts of the test incorporate brief *practice* trials. Here it is especially important to emhasize speed. If the client is overly meticulous, wait until the sample has been finished, and say:

> *Remember that you get more points for working quickly and you can go really fast* [demonstrate by joining a few circles with rapid pencil strokes]. *You see, it's okay to draw the lines very quickly. Let's try that again on this one* [hand client a clean sample]. *Now show me how fast you can draw the lines.*

A useful instruction formula on the practice items for Part B is as follows:

> *On this page are some numbers and some letters. Begin at number 1* (point) *and draw a line from 1 to A* (point), *from A to 2* (point to 2), *2 to B* (examiner continues to point to the appropriate letter or number), *B to 3, 3 to C, and so on, in order, until you reach the end, here* (point). *Remember, first you have a number, then a letter, then a number, then a letter, and so on. Draw the lines as fast as you can. Ready? Begin!*

If necessary, provide another practice sheet and repeat the speediness instructions given for Part A.

***Mazes.*** We have for many years successfully used this WISC–III test with non-test-wise adults and children, who, if unimpaired, turn in scores at or above the U.S. norm for age (for adults, the norms for 17-year-olds are used). The instructions as given in the WISC–III manual should be used, paying special attention to the procedure on the practice item: Note that the examiner does not enter the blind alley when pointing to the error—to do so would "give permission" to the client to do likewise. You should stress the need to plan, and also that there is a time limit, albeit generous.

The task demands are readily understood, and there is no need for practice beyond that contained in the first few very easy test items.

***Tower of London.*** This and the New Tower of London tests are perennial favorites with clients because they have the feel of games rather than tests; and, because they look like games, clients perceive them as less novel, and therefore less threatening, than many other items in the battery. The difficulty gradient is gradual, and there is no need for additional practice items. The administration and scoring methods that we have found to be most informative are given in Appendix 12.

However, as Hanes (1996) pointed out, bright clients with subtle planning impairments do not find even the five move items especially taxing. In such cases, the New Tower of London is the instrument of choice. There are 4 beads instead of 3, 16 items instead of 10, with the most difficult requiring nine moves to the solution instead of the maximum of five on the Shallice version. This is a copyrighted test, and ordering information is given in Appendix 4.

***Austin Maze Test.*** For non-test-wise clients with 12 or more years of education, this is a useful method for the examination of orderly learning with effective error utilization; indeed, with regard to perseveration and failure to maintain set, the information yield is directly comparable to that from the Wisconsin Card Sorting Test. Walsh (1985, pp. 236–237) described the apparatus, administration procedure, and qualitative interpretation.

## Language

Neuropsychologists are not speech pathologists, but language competence must be screened both to determine if a specialist referral is necessary, and to exercise due diligence, lest a clean bill of health be given to an individual whose reading or writing has been significantly compromised.

***Controlled Oral Word Generation.*** We have for many years used two parallel versions of this test: by letter of the alphabet[9] for those with 12 years education or more, and by category (animals, fruit and vegetables, and articles of clothing) for illiterate or semiliterate clients. These tests are readily understood and accepted by all clients, although interpretation must be cautious unless local norms are available.

---

[9]Letter frequency for the language in question is readily determined by counting dictionary pages for each letter, and selecting the three letters that occupy the largest number of pages, which will ipso facto be those with the highest frequency.

***Telephone Numbers to Dictation.*** Seven-digit telephone numbers are now a worldwide standard (10-digit numbers are the norm for mobile telephones) and we routinely include a telephone number transcription test for all except illiterate clients, because among urbanized people, as noted previously, numeracy is near-universal. We give the 10 numbers listed in Appendix 13. Anything less than 8 or 9 correct raises the suspicion of an attentional or auditory processing deficit.

***Words and Sentences to Dictation.*** Again, regardless of educational level, clients whose occupational history indicates they had been able to write with reasonable accuracy should be given one and two syllable words and a sentence in their home language. Writing should be evaluated for accuracy and legibility.

***Reading.*** The previous reasoning again applies: If the occupational or educational history suggests that clients were able to read at a certain level, test fluency and accuracy by asking them to read aloud a simple passage in their home language. (We use one of the memory stories in our repertoire after narrative memory testing has been completed.)

The *Token Test* is a useful quick screening method for the detection of receptive language difficulties. Give two practice items emphasizing that the client must attend closely because instructions cannot be repeated. In the brief 16-item version of the test, only one command (Item 12: "Touch the blue circle with the red square") contains an object–agent construction. If there is an error or a hesitation on this item, probe further after the test has been completed by placing, for example, scissors and a ruler on the table, and making up your own commands, such as "Point the ruler at the scissors" (here agent and object are not reversed). Let a pause of interpolated testing intervene, then say, "Now point to the ruler with the scissors," which does reverse the sequence.

*Chapter* **11**

# Structuring Behavioral Reports

How can the mass of demographic information, behavioral description, and test results now be woven together to make a vivid and accurate picture of your client, and above all a picture that will answer the questions that your referral source has put to you?

These questions may be narrow and specific (e.g., if an ophthalmologist needs more information about the reading difficulties of a brain injured client), or broad and wide-ranging (e.g., a psychiatrist asks about personality functioning, dangerousness, and optimum treatment). If a rehabilitation team makes the referral, they may relate to the capacity to learn and retain new information in a variety of different modalities. A report for the primary caregivers and extended family will deal with prognostic and day-to-day management issues.

To complicate matters even further, the "discourse register" of the report—to borrow a term from chapter 7—is very different for health care professionals on the one hand and lay readers on the other. Of course, even for professionals, it is always a good idea to explain in a sentence or two the content of each of the tests and what it measures: You cannot expect neurosurgeons or physiatrists to be familiar with Trails A and B or the AVLT (for further suggestions on this point, see "Test Performance" later). For caregivers and the family, technical jargon should be replaced by simple English without talking down to your readers.

As a result, reports vary enormously in focus, length, vocabulary, tone, and the stringency of the argumentation that leads to the given conclusions. What I have chosen to do is to describe the structure and content of a forensic report, which is more comprehensive, more lucidly written, and more stringently argued—because it will be more rigorously tested—than most other types of report. Using this model has another advantage: In the British legal system and its derivatives in other countries, all reports seek to determine what the client could do before the accident or injury and what they can do now. In the courtroom, this comes down to a very specific quantitative question: What is the client's future loss of earnings? If you can write a forensic report that answers that question sharply enough to convince a judge and jury, then you will be able to write coherent and well-reasoned reports for other referral sources (in this context, see South African Clinical Neuropsychology Association, 1995).

## HISTORICAL AND BEHAVIORAL REVIEW

### Identifying Information

The report begins with tabulated demographic information that serves as a quick ready-reference summary: the client's name, date of birth, and present age, the date of the incident that gave rise to the brain damage (from here on, I usually refer to "the accident," because most forensic referrals relate to motor vehicle accidents), the interview and test date, the persons present at the interview, the documentation you have available (if a lot of it is irrelevant, you may prefer to specify only the documents you have consulted), and, finally, the referral source.

### Background

This is the first narrative section of the report. Here, mindful of the adage that apples do not fall far from the tree (chap. 4), you deal with the client's intellectual and social endowment, and their present family circumstances. Next, you describe your client's education. In general, the younger the person, the more emphasis you place on the current educational context because that is all the evidence you have with regard to premorbid ability and what the client might have achieved but for the accident. As people mature and enter the world of work, education recedes from foreground to background, with relatively greater emphasis given to workplace achievements under a heading such as "Employment" or "Career." Note that up to this point, all the information you have given predates the accident or illness.

### The Accident

Unless you have been retained as a malpractice consultant, the emphasis in this section is on the *consequences* of the incident and not on the *circumstances*. The cause of the incident—who was on the wrong side of the road or who botched the anesthetic—is contested ground, and the less said about fault and liability, the better. For neuropsychological purposes, it is the nature and severity of the brain damage that is important. The exception to this general rule is that the physics of the injury (i.e., the nature of the forces applied to the head) determine the nature of the injury. A heavy blow to the unsupported head with a blunt weapon (e.g., a baseball bat) gives rise to coup–contre coup injuries. A heavy fall onto the occiput is likely to give rise to cavitation effects at the frontal poles, as noted in chapter 7. In motor vehicle accidents, the position of the victim and the mechanics of the accident give rise to different kinds and severity of injury, so it is helpful to say, without assigning blame, whether your client was a driver, passenger, or pedestrian; whether it was a head-on or rear-end collision; whether the vehicle overturned; and so on.

A useful sequence is to move now to the client's own subjective account of the incident. Record here the time and content of the last memory before the accident, which determines the duration of retrograde amnesia. However, as noted in chapter 7 ("Posttraumatic Amnesia: The Puzzle of Variable Duration"), the first memory after the accident does not necessarily mark the end of posttraumatic amnesia.

Deal now with the postincident history (how soon the client was stabilized and reached hospital) and the medical data—the initial Glasgow Coma Scale, the hospital assessment of the severity of the brain damage, and the course of treatment, with special emphasis on incidents that compromise CNS integrity (e.g., cerebral hypoperfusion arising for example from fat emboli, respiratory failure, or cardiac arrest).

The most important part of this section is your assessment of the severity of the brain injury, which as noted in chapter 7 is determined by the depth and duration of unconsciousness, and the duration of retrograde and posttraumatic amnesia. With regard to the former, neither neurophysical normality, or a Glasgow Coma Scale of 15/15 on admission (Nell, 1997b), can rule out the existence of subtle but devastating cognitive and behavioral deficits (Benton, 1989; Blakely & Harrington, 1993; Leininger et al., 1990; L. F. Marshall & Ruff, 1989). For the reasons noted in chapter 7, the neuropsychologist cannot shirk the duty of determining injury severity even if this means going against medical opinion.

### Postinjury Education and Employment

A *watershed event* is one that gives rise, from that time on, to a discernible change in the client's abilities and behavior at home, in the classroom, or at work. The question that lies at the heart of the behavioral method is whether the accident or illness constitutes a watershed in the client's life. This is the first of three places in the report that address the question of the watershed: The others are the behavioral information recorded under "Other Complaints," and the client's test performance. These three sections lay the basis for the conclusions that you will draw under "Formulation and Prognosis."

**Presentation**

Record here what you see, and not what you think: The Formulation section is the place for your own conclusions. Your purpose is to paint a vivid picture of your client, and when appropriate the caregivers, so that the reader will see in mind's eye who it is that you are talking about. There is no need to comment on the unremarkable, but there is much that may seem unremarkable that tells a story:

> Mr. Selipe is a man of above-average height, smiling and friendly, who was at once relaxed with the examiners. He was well groomed and colourfully dressed in clean, well-pressed clothing, but wore unmatching socks; his shoelaces were untied. He answered at length and discursively when a direct question was put to him, but after the first few minutes of the interview, he showed little interest in the proceedings, even though these concerned his own history and current behaviour, most information coming from his mother and elder brother. While the interview was in progress, he yawned, fell asleep briefly, and then wandered around the consulting room, moving ornaments to new positions and inspecting books. He smiled and laughed a great deal, even when talking about distressing events such as the death of his best friend in the accident.

Why say his personal hygiene was good and that his clothing is clean? Such details are surely unremarkable. But here the examiner has described a disinhibited man who is socially inappropriate, behaving in a professional's consulting room as if in his own home. He is also hyperverbal and emotionally inappropriate. Given this level of hypofrontality, it is indeed remarkable that he is well-groomed, and therein lies a clinical question: Who is the devoted caregiver who insures that this man is well cared for? Because, if left to his own devices, he obviously cannot be bothered to select matching socks or tie his shoelaces. This section draws no conclusions, but in the mind of the experienced reader it raises the possibility that disinhibition with poor temper control are present, perhaps together with a flightiness that makes it difficult for him to maintain set for more than a few minutes at a time.

The stage is thus set for reevaluation of these as-yet-unspoken hypotheses, which will be either confirmed or rejected by the behavioral information and the test results.

## Main Complaints

It is helpful to structure reports in parallel with the natural progression of the interview. Clients find it easier (and less threatening) to begin with the physical difficulties (pain, weakness, sleep patterns, fatigue) and move on from there to readily perceived cognitive difficulties such as memory, and then to related problems such as orderly thinking and problem solving. Changes in personality are the most sensitive area, and therefore are best left until later in the interview. A report structured in the same sequence reads logically and is easy to follow, even for lay readers unfamiliar with the inner logic of the concussion syndrome.

***Physical Difficulties.*** The purpose of this section is not to recapitulate the details of the orthopedic, neurological, or occupational therapy findings, but to lay the groundwork for the conclusions you will later draw about your client's work capacity and career potential. Your recommendations for physical rehabilitation and assistive devices (with deference here to the occupational and physical therapists) flow from this section.

*Arousal level* provides a natural transition from the physical to the behavioral-cognitive domain. "Fatigue" is often useful as the next heading, commenting in particular on hours of sleep and daytime weariness. Alcohol tolerance, invariably reduced by hypoarousal, can come next, followed by what your client has told you about attention and concentration, which are both undermined by a lower level of arousal.

***Cognitive Changes.*** There is now a natural transition in the report from arousal and concentration to related cognitive problems, beginning with memory difficulties—which are the most common change in thinking and the most pervasive in their effects. Move on from there to problem solving in its various modalities: troubleshooting, mental arithmetic, making change, and so forth. Apraxias and agnosias are uncommon after closed head injuries, but if present should be described here. The most likely manifestation is as visuospatial deficits, which will find expression in complaints about making sense of diagrams and in problems with orientation in space.

***Changes in Personality.*** A convenient entry point to this sensitive area is bad temper, which you can usefully conceptualize as a problem of social appropriateness, moving on from there to other socially disruptive behaviors that are potentiated by a reduced level of arousal (i.e., hyperverbality, flightiness in speech and in task performance, and poor self-monitoring). Under each of these heads, begin with the client's own views, go on to the caregivers' accounts, and then deal with the spread: Do these behaviors manifest only at home, or also in public and work settings? Disinhibition in any of these areas corrodes relationships, so it is natural to deal next with affectionate and working relationships.

There is a natural movement from this discussion first to the vividness and appropriateness of *emotionality*, which (as noted in chap. 7) is a principal motivational driver, and then to *mood*. Reactive depression is a common consequence of brain injury, which may have been associated with a near death experience, and this is the appropriate place to deal with it.

***Tell a Story.*** Behavior is a real series of events in the real world, and the best way to describe it is first to do it conceptually and then by way of a vivid recent example. For example, "The greatly reduced level of arousal has given rise to a significant rage dyscontrol syndrome, from which the client derives emotional release as well as a heightened sense of personal control." A concrete example then follows:

> Mr. Schoonveldt's wife says that the day before the present interview, as she was getting the children ready for school, she heard a commotion from the bedroom. She found her husband smashing the plaster cast on his right arm against the wall and the doorframe, screaming the while that he was not a cripple and no-one was going to turn him into one. Mr. Schoonveldt listened attentively and with evident approval to this account, then explained that he had been attempting to tie his shoelaces and had been unable to do so because of the cast.

The immediacy of this description tells a great deal more than the conceptual formulation on its own.

### Behavior During Testing

As with presentation, describe here what you see (rather than what you think) as it reflects on the client's attention, effort, and motivation. Do clients attend fully to instructions and wait for these to be completed before starting, or rush into the test while the examiner is still explaining what has to be done? Do clients understand the instructions or do they have to be explained more fully? Do clients remember instructions and problems, or do they have to be repeated? Do clients offer spontaneous help in packing away equipment? Do clients engage appropriately with the examiner, and stay on task even when the examiner's attention is withdrawn? Does fatigue manifest as the test session draws on?

Finally, are clients making an effort to do well and turning in the best performances of which they are capable? If so, it is important that you validate your test results by confirming this. If there is any doubt on this score, you should say so here, and go on under the next heading to examine this possibility.

### Role Enactment and Malingering

There is an important distinction between *malingering*—deliberately doing badly on tests or feigning symptoms in order to gain compensation—and *role enactment*, in which the individual is convinced that certain deficits are present, then presenting complaints and test performances that confirm these problems. I have yet to see malingering in clients from non-Western backgrounds. On the other hand, malingering is in great diagnostic favor with defendant experts who find this to be the most attractive explanation for the markedly depressed test performances often seen after mild brain injuries. Indeed, to paraphrase what my postdoctoral supervisor, the late Chuck Matthews, said of anxiety as an explanation for poor test performance, the charge of malingering is often the last resort of scoundrels! So, unless you have hard evidence of role enactment or malingering, you should be hesitant about making this attribution.

There is now a large literature on the detection of malingering, including two recent books (Reynolds, 1998; Rogers, 1997), and a plethora of new or revised methods (Chouinard & Rouleau, 1997; Cochrane, Baker, & Meudell, 1998; Griffin, Glassmire, Henderson, & McCann, 1997; Rees, Tombaugh, Gansler, & Moczynski, 1998), several of which may prove suitable for use with non-test-wise clients.

This concludes the historical and behavioral sections of the report. Next the discussion moves on to the test results, relating these at every step to real-world problems and behavior.

## TEST PERFORMANCE

I invariably report tests in the sequence in which they appear in Table 10.2—in other words, from simple to more complex tasks, or, in information-processing terms, from performance processes to metaprocesses. With regard to sequencing, therefore, this section arranges itself. But the difference between a competent report and an excellent one is in the subtle links that can be forged between seemingly unrelated aspects of test performance: between digit span forward and initial word span on a list learning task, between errors on mazes and impulsivity on the Tower of London, or between visuospatial difficulties and unexpected breakdowns in mental arithmetic. Even more interesting are links between clients' test performances and behavior: for example, postaccident inability to troubleshoot as a machine operator and a chaotic Austin Maze performance, criminal charges of shoplifting and severely impaired narrative memory, or daytime fatigue and catnapping with steadily declining test workrate and accuracy.

### Explanatory Appendix

As already noted, in today's rapidly changing markets for neuropsychological assessments, you cannot assume that even your professional readers are sufficiently familiar with the test repertoire to know what the client's task is on Digit Symbol Substitution or the AVLT, and what meaning attaches to a good or a poor score. However, reports become painfully long if these large boluses of explanatory information are a fixed feature. One answer is to produce a substantial appendix that is both descriptive, explaining the client's task on each test, and interpretive, saying what high or low scores typically mean. Fortunately for the trees, you need to supply this appendix only once to each of your referral sources.

### Formulation

*Formulation* is a term that was much favored in clinical reports earlier in the century, but has since dropped out of use. I retain it as a heading because it neatly captures the purpose of this section, which is to formulate concisely the meaning of your findings, that is, to weave together the many different inferences and hypotheses that have arisen in the preceding material. In this section you at last draw conclusions, and show how a central trunk of fundamental changes in your client's arousal, cognition, and personality gives rise to a multitude of linked behaviors, and how these changes are reflected in the test results. You must, however, remember that the formulation is not a summary but an argumentation. If your readers need a summary, provide one by all means, but do it under that heading and preferably before the formulation.

### Prognosis

It is often helpful to write this section as a set of contrasts. Given the family background and achievements up to the time of the brain damage, what would this person's future prospects have been but for the accident? And, in the light of the conclusions reached in the formulation, what are they now? Will the client be able to complete primary school, high school, or get a university degree? Will the client retain open market employment, and if so, what are the promotional prospects? Finally, given the treatment you will recommend in the next section, to what extent are these prospects likely to be improved?

### Recommendations

This final section is first a summary review of recommendations you have already made for workups by other disciplines such as orthopedics, neurology, speech therapy, occupational therapy, and so on. Then, in the light of the previous section, you set out your client's needs for specialized therapies, in particular neuropsychological rehabilitation and psychotherapy, specifying if possible the modality (e.g., cognitive behavior therapy) that is likely to be most effective. Forensic reports also need to consider the client's competence to handle a financial award in their own and their family's best long-term interests.

## INTERPRETING THE TEST SCORES OF NON-TEST-WISE CLIENTS

In the absence of reliable norms, there are a number of interpretive avenues open to the neuropsychologist working with non-test-wise clients.

### Comparative Interpretation

If a school or work record is available, test results can be compared with work requirements in order to draw conclusions.

> A butcher's van driver with 5 years education sustained a brain injury. He had held the job for 7 years, and his duties were to use the delivery dockets to plan a daily route, and ensure that each customer confirmed receipt of their order. He was dismissed soon after his return to work. On testing, he was unable to transcribe telephone numbers to dictation, made 5 errors on the brief Token Test, and on the WISC–R, he had standard scores of 5 on the Arithmetic Subtest and 6 on Mazes. Profound visuospatial difficulties consistent with his depleted performance on Mazes emerged on the CFT and Draw a Bicycle tests.

Would it be scientifically responsible to report that this formerly competent man had standard scores of 5 on the Arithmetic subtest of the WISC–R and 6 on Mazes? After all, the construct and predictive validity of these norms are unknown. It is responsible if you clearly state in the report that the norms are for 17-year-old American children, and if you are able to validate your interpretation of the score on ecological grounds by saying that his test performances were incompatible with his former level of employment, and that the brain injury was causal. To this you would add that he was incapacitated for work requiring orientation in space, such as planning and executing a delivery route, and would be unable to cope with alternative employment requiring intact receptive language skills. This intraindividual comparison method is often the most direct and powerful interpretative strategy.

In your report, you might thus deal with the norms issue along the following lines: "Although these norms are inappropriate for an adult South African (or Nigerian or Nicaraguan), it is noted that the items the client failed required no more than simple addition or subtraction that in terms of the work history should have been well within his grasp."

***Placement and Educational Tests.*** Even in underdeveloped settings, industry makes insistent demands for locally validated personnel selection and placement instruments; similarly, the assessment needs of the school system will often be addressed by a range of tests. For whatever reason, the clinical and neuropsychological communities are often unaware or disdainful of these individual and educational tests, which on the contrary make an invaluable and often well-validated contribution to the neuropsychological armamentarium.

***Preschool Brain Injuries.*** For children injured before they have begun school or before a stable school record has been established, the intraindividual comparative route is trickier but still viable. There are in the first place the developmental milestones, supplemented with reports by the parents, siblings, and perhaps neighbors that the child was normal. However, this is weak courtroom information because it rests on the subjective judgment of interested parties. Sibling comparisons have something of the same weakness, but may nonetheless be stronger, for example, that the milestones and self-care abilities of the injured child had outstripped a sibling for whom school records are available.

Indeed, the only thing better than local norms is an identical twin! We were recently lucky enough to have just such a case in our practice. A comparison of the injured child with her twin, who was assessed at the same time for comparative purposes, vividly demonstrated the typical concussional pattern of lowered levels of arousal; fluctuating attention span; difficulties with concentration, sequencing, and memory; psychomotor slowing, with considerable frontal involvement in the form of adynamia; low motivation; poor impulse and temper control; and poor self-monitoring.

Here is the case as it was written up for presentation by Marilyn Adan, a Johannesburg neuropsychologist:

> Phindi and Dudu are identical twins resident in Alexander Township in Johannesburg. At age 4, Phindi was struck by a car and and transferred from a local clinic to the Johannesburg Hospital. Some four hours post-accident, she was mute and incontinent with a Glasgow Coma Scale of 8. A CT scan showed bilateral parieto-occipital contusion with haematomas.

On interview 4 years later, the mother said that Phindi had been "the clever one," but post-accident she failed Grade 1 with a 29% aggregate compared with Duduls 75% pass mark. During the interview with their mother, the twins' facial expression and general behaviour differed markedly. Phindi appeared glum and dull-eyed, showing few changes in facial expression. She sat quietly during the initial stages of the interview. Dudu, on the other hand, was bright-eyed and demonstrated a full range of facial expressions. She looked around the room, showing an interest in the ornaments and pictures on the walls. She also fidgeted and moved around in the chair within normal limits for a child of her age.

Dudu was the first to examine a box of toys offered to the children. She selected a pack of cards and played with them for about 15 minutes, examining them carefully before sorting them according to commonalities. Thereafter, she drew pictures and wrote words with the felt-tipped pens and pencils provided before returning to play with the cards. Throughout this time, she chatted excitedly to her sister and was obviously enjoying the games she played.

Phindi watched her sister and followed her example by playing with another pack of cards. However, there appeared to be no real purpose to her game and she soon lost interest, alternating between handling a teddy bear, drawing and playing with the cards. She seldom initiated conversation, but responded when her sister spoke to her. After about an hour, she lay down and showed no further interest in the toys, unlike Dudu who continued to amuse herself throughout.

During the testing their behaviour differed significantly. Phindi appeared withdrawn and adynamic. She tired easily and required short breaks during which she slept. Her attention span and concentration fluctuated and, although she persisted with tasks, she seemed generally lethargic and unmotivated. Errors were neither spotted nor corrected and she did not offer to assist in packing away test material. She worked slowly and handled test material clumsily.

By contrast, Dudu appeared alert and eager to start working on each new test that was presented to her. There was no sign of fatigue and she offered to assist in packing away the test material. She was also motivated by success and praise. Her attention span and concentration were within normal limits for a child of her age. She was aware of errors she made, but did not always correct them and her work speed fluctuated.

Both girls were asked to tie their shoelaces and fasten their buttons. Dudu coped well, while Phindi fastened her buttons slowly and could not tie her laces without assistance.

Given the WISC–R Coding Subtest, Dudu managed 16 substitutions, a standard deviation better than Phindi's 10. Phindi's slow performance on Coding is in keeping with the difficulties she experiences in copying from the board in class.

On the SSAIS–R Digit Span Subtest, both girls recalled 4 digits forwards but Dudu 3 backwards and Phindi only 2. However, the quality of their performance differed. Dudu's scaled score was 9 but Phindi's only 3.

> Dudu counted forwards and backwards from 20 with no difficulty but Phindi made three erors when counting forwards and was incapable of counting backwards, saying 20, 19, 18, 16, 19, 15, 40, 100. On the Luria Echopraxia Tapping Test, Dudu self-corrected but Phindi was unaware of her errors.
>
> On the Vocabulary Subtest from the Individual Scale for Zulu Pupils, standardised for children of 9 and older, Dudu's scaled score was 7, which is good considering the age difference between herself and the norm group, but Phindi's score was only 3. Again, she gave up easily and required encouragement to persist.
>
> Phindi wrote her name with letter transpositions and reversals, could not recite the alphabet, or identify letters. Dudu on the other hand wrote her name, letters to dictation and some simple Zulu words correctly. She could also recite the alphabet. On Phindi's performance on the WISC–R Mazes Subtest, much impulsivity and poor planning were evident resulting in a scaled score of 5 on the the American norms. Dudu's score by contrast was 10. (Adan, 1997)

Unfortunately, this serendipitous monozygotic circumstance is rare, and brings no release from the duty to establish valid norms. On the other hand, and fortunately, this is an endlessly fascinating and rewarding task, and I feel privileged to have given it a large part of my working life as a psychologist. My last word, in the penultimate sentence of this monograph on how to test without norms, is to hope that it will within a decade or two have become redundant. To paraphrase Freud, "Where chaos was, there shall order be."

# APPENDICES

*Appendix* 1

# Piaget's Stages of Cognitive Development

Some familiarity with Piaget's classification of cognitive stages is needed to follow the piagetian experiments in chapter 3. The following brief reprise is helpful for readers who have forgotten their Psych 101.

## CONCRETE AND FORMAL OPERATIONS

From birth to age 2 is the period of *sensorimotor intelligence* (Stage 1). The child makes simple perceptual and motor adjustments to objects in a gradually more differentiated world, but does not carry out symbolic manipulations of these objects. From age 2 to 7 is a period of *preoperational* representations (Stage 2A), in which representational thought begins and is elaborated, coming to full expression in the period of *concrete operations* from age 7 to 11 (Stage 2B). By operation, Piaget understands a representational act that forms part of a strongly structured representational totality: Indeed, the centrality of mental representations is one of the two core features that define modern cognitive science (Gardner, 1987, p. 38).

Conservation of quantity is an example of a concrete operation. A child entering this stage begins to recognize that when water is poured from a wide shallow container into a tall narrow one, the shape of the liquid has changed, but the quantity has been conserved. However, the preoperational child centers attention on the end state of the transformation, is overimpressed with the great height or width of the fluid, and makes the error of saying either that there is now more liquid or less in the container: "A single, isolated cognition of this sort, with little or no systematic reference to other cognitions past or potential, is the hallmark of the preoperational child" (Flavell, 1963, p. 167). However, the Stage 2 child, able to carry out concrete operations, sees a particular transformation as just one example of any number of possible transformations in which the height, width, shape, or orientation of the liquid can vary singly or together without affecting the quantity of liquid. The older concrete operational child does something with which the younger preoperational does not:

> He brings to bear a whole system of potential operations on the specifics at hand, and, by so doing, can see each specific, not as an ultimate, but as the "is" of a "could be" totality, other specifics seen as being at any time substitutable for it. (p. 168)

From age 11 to 15 is the period of *formal operations* (Stage 3) in which a final cognitive reorganization takes place. This enables children to deal effectively not only with the reality that confronts them, which the concrete operational child does perfectly well, but—citing Flavell (1963), on whose work the preceding summary of Piaget's cognitive stages is based—also with

> the world of pure possibility, the world of abstract, propositional statements, the world of "as if." This kind of cognition . . . is adult thought in the sense that these are the structures within which adults operate when they are at their cognitive best, that is, when they are thinking logically and abstractly. (p. xx)

Formal operations are "the crowning achievement of intellectual development, the final equilibrium state towards which intellectual evolution has been moving since infancy" (Flavell, 1963, p. 202), and is distinguished from concrete operations because it contrasts the real here and now that is present to the senses with the possible: This is the essence of the "abstract attitude" first described by Kurt Goldstein in the 1930s. Neuropsychologists will not need the reminder Flavell gave his readers that "there is nothing trivial about this reversal in role; it amounts to a fundamental reorientation towards cognitive problems" (p. 205). The child at the level of formal operations is no longer preoccupied with what is present, but can turn to discovering "the real among the possible," and proceeds through propositional thinking to operate on the results of previous operations; this reflexivity, this thinking about thought, led Piaget to call formal operations "second degree or interpropositional operations" (pp. 205–206).

The child's progress through these two stages of concrete and formal operations is beautifully captured in a 1958 book, *The Growth of Logical Thinking from Childhood to Adolescence* (Inhelder & Piaget, 1958). From a series of elegant experiments presented to children at various stages of their cognitive development, I have selected one, "Flexibility and the Operations

Mediating the Separation of Variables," in order to present the child's progress through this developmental course in a way that will help clarify the content of the piagetian tasks used by Dasen and others in exotic settings.

The apparatus for this experiment is illustrated in Fig. A1.1. It consists of a shallow square pan containing water, and a clamp through which six different rods can pass and be clamped at any point along their length. The rods are either thick or thin, round or square, and made of steel or brass. There are three different weights of 100, 200, and 300 grams, each in the form of a manikin that can be attached to the end of a rod. Whether or not the tip of a given rod will touch the water therefore depends on the interaction of five distinct variables: the cross-section of the rod (round or square), the material of which it is made, the rod thickness, the length of the rod (i.e., the position at which it is clamped), and whether a lighter or heavier weight has been attached to it.

The cognitive task facing the child is to separate these variables from one another and to determine the part played by each in getting the rod to touch the water. As Piaget noted in his introduction to this part of the

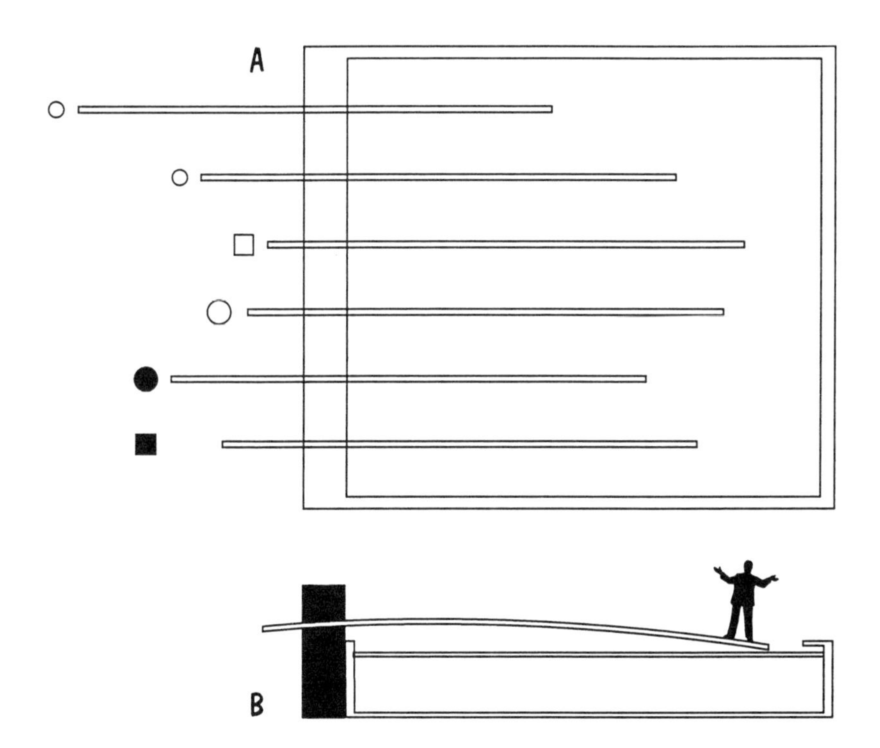


FIG. A1.1. Apparatus to demonstrate concrete and formal operations. Reprinted from Inhelder & Piaget (1958), by permission of Basic Books, a division of Perseus Books, L.L.C.

book, the problem appears at first to be purely concrete, but in fact requires for its solution the use of interpropositional operations that are found only in formal and not in concrete operations.

The top diagram is a plan view of the container, showing the six flexible rods, and, to the left of each rod, its cross-section and thickness. The two lower cross-sections are shaded to indicate that these rods are made of brass. The lower drawing indicates how the rod can be clamped at any point on its length by the block on the left, and how a weight, in the form of a little man, can be attached to the end of the rod, causing it to dip down toward the water; the rod touches the water only when the combination of all five variables is in a specific relation to one another (from Inhelder & Piaget, 1958, p. xx).

Piaget gives transcripts of the conversations of children at the preoperational, the concrete operational, and the formal operations stages to this problem.

### Preoperational Stage

A child who is exactly 5 years old puts 200 g onto the thick square steel bar and says, "It doesn't touch the water." Then he attaches the weight to the thin round brass rod, which does touch the water "because it is lower down." He then tries the thin round steel rod, and says this does not touch "because it's on a plank," ignoring the fact that all the other rods are attached to the same plank: At this age, children are limited to simple reports of what they perceive, and deal in logical classes that are undifferentiated and too general.

### Concrete Operations

A child of 9-2 (9 years and 2 months) says, "Some of them bend more than others because they are lighter (he points to the thinnest) and the others are heavier." Asked to show how the lighter rod can bend more than a heavy one, he places the same weight at the end of both, without noticing that the thin rod he has chosen is also the longest. Then, asked to show the role of length rather than of thickness in bending, he again puts the same weights on the same two rods, now saying that the role of length is demonstrated, but not noticing that he had used the same arrangement a moment before to demonstrate the role of thickness.

### Formal Operations

At first, this stage is characterized by the use of some interpropositional operations, such as implication and exclusion, but the 11- or 12-year-old child cannot organize the necessary systematic proof, which is to change only one variable while keeping the others constant; or, as Piaget put it, solving the problem means the child must use the schema, "All other things being equal." For example, a child 11 years and 10 months old says that the long thin rods will bend most, and when asked to demonstrate this, he chooses two steel rods of the same length and cross-section, but one thicker than the other, puts 100 g on the thick rod, and 200 g on the thin one, saying, "That one bends more." Asked if this is the right way to demonstrate that the thin rod bends more, he replaces the 100 g with a 200 g weight on the thicker rod. In other words, the condition of "all other things being equal" is met only with difficulty. As Piaget noted, there is here a search for equivalence in the conditions of comparison, but the proof is not yet rigorous.

This changes with the acquisition of full command of formal operations, which Piaget designated as the second stage of this third period of development. For example, a young woman of 16 years 10 months is entirely systematic in holding the variables equal, working first with weight, then with cross-section, then with material, then with the length of the rods, refusing to compare rods of different section or material. This schema, noted Piaget, although apparently simple, can only be used spontaneously from age 14 or 15 (from Inhelder & Piaget, 1958, chap. 3).

Appendix 2

# Intake Form

Please fill in this form as carefully and accurately as possible.
Take your time - you'll need at least an hour.

Today's date: ______________________

Surname : ______________________________________________

First names: ______________________________________________

Date of birth: ______________________ Age: ____________

Place of birth: ______________ Home language: ____________

Referred by: ______________________________________________

Questionnaire filled in by : ______________________________

Other persons present at testing:

| Name | Relationship or role |
|---|---|
| | |
| | |
| | |

| Street address of your home | Telephone numbers |
|---|---|
| ______________________ | Client (H) ______________ |
| ______________________ | (W) ______________ |
| ______________________ | Caregiver (H) ______________ |
| ______________________ | (W) ______________ |

## 1. The Accident or Illness

1.1 Date Day of week Time of day

______________________________________________

1.2 Circumstances:

[ ] Pedestrian
[ ] Driver
[ ] Passenger
[ ] Gunshot
[ ] Illness
[ ] Cyclist/motor cyclist

## 2. Home Background

2.1 Your Family
List your family, brothers and sisters from oldest to youngest. Underline your own name. Put step-, half-brothers and sisters in brackets.

2.1.1 Your parents:

| | Name | Age now | Marital Status (S,M,D)* | Highest Standard | Occupation now |
|---|---|---|---|---|---|
| Your father: | | | | | |
| Your mother: | | | | | |

Date your parents married:____________________

2.1.2 Your step parent (if applicable):

____________________

2.1.2 Your brothers and sisters:

| Name | Age now | Marital Status (S,M,D)* | Highest Standard | Occupation now |
|---|---|---|---|---|
| | | | | |
| | | | | |
| | | | | |
| | | | | |
| | | | | |

2.2.1 Your spouse/partner (if applicable):

| Name | Age now | Highest Standard | Occupation now |
|---|---|---|---|
| | | | |

2.2.2 Your previous spouses/partners (if applicable):

| Name | Age now | Highest Standard | Occupation now |
|---|---|---|---|
| | | | |
| | | | |

* S-single; M-married; D-divorced

2.3 Your Children:

| Name | Sex M/F | Age now | Marital Status (S,M,D)* | Highest Standard | Occupation now |
|---|---|---|---|---|---|
| | | | | | |
| | | | | | |
| | | | | | |
| | | | | | |
| | | | | | |

OMIT SECTION 3 - GO TO SECTION 4

## 3. Your developmental/medical history

3.1 Complications - pregnancy, birth, weight, feeding, sleeping:

3.2 Milestones: sit crawl talk walk toilet self-care

Age:

3.3 Social development:

3.4 Previous hospitalization:

3.5 Previous/current medication:

## 4. Military Service

Give dates and details:

### 5. Education and Training

To complete our report we need copies of school/college/technikon/university records. If the accident occurred during the course of education, please ensure that we have copies of pre- and post-accident records.

5.1 Schooling

Fill in the names of all schools you attended

| | Town or district | Year From | To | Highest Standard Passed |
|---|---|---|---|---|
| Primary Schools | | | | |
| | | | | |
| | | | | |
| | | | | |
| | | | | |
| High Schools | | | | |
| | | | | |
| | | | | |
| | | | | |
| | | | | |
| | | | | |

3.2 Standards failed________________________________________

5.2 Post School Training:
Complete this section only if you attended a tertiary educational institution after leaving school.

| Institution Name | Degree/ Diploma | Years attended From - to | Major Subjects | Final year marks |
|---|---|---|---|---|
| | | | | |
| | | | | |
| | | | | |
| | | | | |

## 6. Career

6.1 Pre-accident work record:
List in chronological order all the jobs you held before the accident or illness. State "not working" for the periods you were unemployed and give dates.

| Name of Company | Job Title | Dates from - to | Salary on leaving | Reason for leaving |
|---|---|---|---|---|
| | | | | |
| | | | | |
| | | | | |
| | | | | |

6.2 Post-accident work record:

6.2.1 When did you return to work?____________________

6.2.2 List in chronological order all the jobs you held since the accident or illness. State "not working" for the periods you were unemployed and give dates.

| Name of Company | Job Title | Dates from -to | Salary on leaving | Reason for leaving |
|---|---|---|---|---|
| | | | | |
| | | | | |

6.3 Present job:

____________________

6.4 Relationship with colleagues:

____________________

6.5 If you are not doing the same job as you did before the accident, state your reason/s for changing:

____________________

6.6 If you are unemployed, state the main reason why you are unable to work:

____________________

6.7 Future plans:____________________

## 7. Main Complaints

This section is divided into two areas - physical changes and complaints, and other problems you may be experiencing in your personal and social life, or at work.

7.1 Physical and related issues:
Describe your present physical condition and how your injuries affect your life at present.

____________________________________________

____________________________________________

____________________________________________

____________________________________________

____________________________________________

____________________________________________

____________________________________________

____________________________________________

____________________________________________

____________________________________________

____________________________________________

____________________________________________

What sports and other free time activities did you do before the accident?

____________________________________________

____________________________________________

____________________________________________

What sports and other free time activities do you do now?

____________________________________________

____________________________________________

____________________________________________

7.2 Other Problems:
Apart from the physical problems you listed in the previous section, do you feel different now? Are there changes in your behaviour or personality or problems in your social or work life that you are experiencing? If there are, describe these below:

## 8. Retrograde and Post-traumatic Amnesia

8.1 Describe your last memory before the accident:

______________________________________________

______________________________________________

8.2 Give as best you can the time and date of this memory:

Time: ______________________ Date: ______________________

8.3 Describe your first memory after the accident:

______________________________________________

______________________________________________

8.4 Give as best you can the time and date of this memory:

Time: ______________________ Date: ______________________

8.5 On what day/date after the accident did you first recognise members of your family?

Day: ______________________ Date: ______________________

How did you acknowledge their presence (smiling, squeezing a hand, or the like)

______________________________________________

______________________________________________

8.6 On what day/date did you know for the first time where you were and what had happened to you:
Day: ______________________ Date: ______________________

8.7 Explain why you give this answer:

______________________________________________

______________________________________________

8.8 From what date can you remember from one day to the next the main things that happened to you and what you did?

______________________________________________

8.9 Explain or give reasons why you think this date is correct:

______________________________________________

______________________________________________

8.10 Hospital information:

Name of hospital Admission date Discharge date

_______________________________________________

_______________________________________________

_______________________________________________

8.11 To be completed by a family member or caregiver who saw the person soon after their accident or illness.

At what time/date did you first see the person?

_______________________________________________

Describe this person's condition at that time - what did you see?

_______________________________________________

_______________________________________________

_______________________________________________

What was this person's condition like after discharge from hospital?

_______________________________________________

_______________________________________________

_______________________________________________

8.12 Apart from the accident you have just told us about, has there been any other incident, before or since, in which you were dazed or unconscious? [ ] Yes [ ] No

If yes, explain:

_______________________________________________

_______________________________________________

_______________________________________________

*Appendix* **3**

# Educating the Executive

How might the process of test development take account of cultural worlds that are very different from those of the test constructor? What follows is the outline of a deliberately unstructured method of inquiry, written as a dialogue between test-maker and client: It is a method of educating the test-maker's executive in order to enhance the cultural appropriateness of the resultant test.

## GETTING INTO YOUR CLIENT'S WORLD

The most powerful way to get into your client's world, and for your client to understand yours, is to create within the assessment dyad the intensity of a child's learning experience with its parents, a magical zone of proximal development (chap. 8) in which, by conversation and demonstration, you both come to terms with what you want (your test as you give it) and what your client understands you to want (your test as perceived). Working in this modality may not be easy for psychologists, who have traditionally been among the less well-mannered social scientists, unpacking their carpet-bag of tests in a stranger's home without waiting for the teacups to cool. Instead, like anthropologists and oral historians, we need to join with our clients' world before inviting them into ours. In this way, the client's executive is more likely to become accessible to us from within its own web.

Suppose you wanted to develop a set of instructions for the Digit Symbol subtest of the WAIS–III. Your conversation with the client might begin by putting your need to administer the test in context, for example, by saying that all people do things differently even if the task is the same; you want to give a lot of people the same test as a way of finding out how many different ways there are of doing the same thing.

Before moving on to the test, you need to establish if your client is sufficiently numerate to identify the numbers 1–9 in 48 point type (Fig. A3.1), and has the visual acuity needed for the tests you plan to administer. If these for example include Digit Symbol and Symbol Search, adequate identification of 30 point numbers will suffice; but if you plan to administer Pursuit Aiming, good 12 point vision is needed.

First give clients a rationale for the test by saying, for example, that not everyone has the opportunity to learn to write. How might it be possible then to find out how well clients would learn to write if they were given the chance? It would be by asking them to copy shapes that were not letters or numbers, but, like letters, needed to be quickly done and recognizably like the model. You can now show your clients the symbols on the Digit Symbol test sheet, and ask them to copy these onto a sheet you have ruled up in squares that match the size of the test sheet. Check that they hold the pencil comfortably and reproduce the symbols in the same size and shape, and let this practice continue until clients have gained some fluency in copying the shapes.

Now you need to explore the issue of care and quickness. How does your client feel about doing things quickly? Is it possible to be both accurate and quick as you have requested? Or, does this make the client uncomfortable, as if asked to do two contradictory things? The client's responses need to be considered and the suggestion made that speed and accuracy

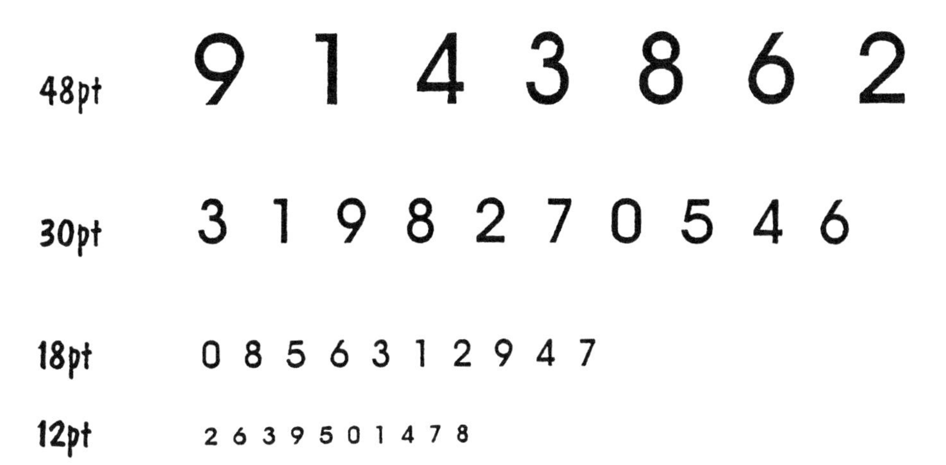


FIG. A3.1. Visual acuity and number identification.

can be traded off against one another in such a way that the largest number of acceptable items is produced in the available time rather than a smaller number of perfect items.

Now introduce the first timed trial on a blank test sheet, encouraging clients to work quickly and accurately, and after each trial to verbalize what went through their head: what had been easy, what was difficult, how they think they have done, and so on. With more articulate clients, it is useful to encourage such verbalization even while the timed trial is in progress. Continue giving additional timed trials, resisting the temptation to stop before clients have reached a performance plateau. Ten successive trials are likely to be a minimum, and more are desirable. After each trial, pause for reflection and comment. The final loop in the procedure is again reflexive: Ask clients what they think you found out about them, what they found out about you and your needs, and about themselves, and to what use they think you might put the information you have gained.

You should ideally repeat this procedure, refining and adapting it with each repetition, with 10 or more clients representative of your target population.

### Instructional Needs

You will have made copious notes during the trial sessions, and now you can systematically list the questions that arose in your dialogues with clients about why you are administering the test, how you and they understand what you want, how this understanding changed in the light of the ongoing dialogue, and so on. This will indicate how the test should be contextualized, what everyone seemed to find easy and can be disposed of in a few words, and which task demands gave rise to difficulties that must be extensively dealt with.

### Practice Needs

As a quick guide to practice needs, you can plot each pilot test client's performance on a graph, with trials shown on the $x$ axis, and number of test items correctly completed on the $y$ axis. Asymptote for each client can be visually determined, and a mean asymptote for the group and its standard deviation computed. The number of practice trials allowed in test administration should be at least equal to the mean, and preferably the mean plus a standard deviation.

As a final step, it might be helpful to systematize the test instructions as three blocks—creating a context, an instructional set, and structured practice that maximizes performance for the final timed trial.

*Appendix* 4

# Test Suppliers

## WHO–NCTB TESTS

The Benton Visual Retention Test (Recognition Form), the Santa Ana Pegboard, and the Pursuit Aiming Test can be ordered through the World Health Organization Division of Occupational Health, Geneva, Switzerland.

An alternative source for the Santa Ana is:

Vaino Levy
Insinooritoimisto Levy Ky
Oskarinkuja 3
02970
Espoo, Finland

## TERRY REACTION TIME APPARATUS

This is obtainable from:

Software Science
3750 Roundbottom Road
Cincinnati, OH
45244 USA
e-mail: information@softwarescience.com

**NEW TOWER OF LONDON TEST**

The publisher is:

Australian Center for Cognitive/Behavioral Treatment and Assessment
P.O. Box 1193
Carlton, Victoria
Australia 3053

*Appendix* 5

# Administration Instructions for the Santa Ana Pegboard

The test consists of a base plate with 48 square depressions and 48 pegs with a cylindrical top and square base. Each peg top is colored half red and half white. The client's task is to turn each peg 180 degrees, as fast as possible. The dominant and nondominant hand are timed separately.

The board with all the pegs inserted into the base plate is placed on the table in front of the client, who must be standing. Make sure that all the pegs are in the same position, white half facing toward the client and red half facing away. Stand next to the client, lift one of the pegs from its depression, and say:

> *Now look at this board with the pegs. The pegs have a round top, but the base is square so that the pegs fit into the holes in the board. Now, your job is to take each peg, turn it around, and put it back, like this.*

## PRACTICE

Demonstrate by turning four pegs, beginning from the left top row. Then replace the four pegs in their original orientation. Say:

> *Now you try that. Lift and turn these pegs with your* [dominant] *hand like I did. Make sure that you are standing in a comfortable position to do it easily.*

Allow 6 pegs for practice and the next 6 for speeded practice with the dominant hand, and the the same for the nondominant hand. Coach the client to work as fast as possible. When the practice is over ask the client to help you turn the pegs back for the test proper.

### Trial 1

Say:

> *Now let's do that while I time you to see how fast you can go. Turn the pegs as fast as possible, with your* [dominant] *hand, beginning from here* [the client's upper left corner].

*Note:* If the client is left dominant, begin in the upper right corner.

> *When you have finished the top row, continue along the next row in the opposite direction, like this* [examiner demonstrates the S-shaped course the client should follow] *and so on until I ask you to stop after. The test lasts only 30 seconds. Please begin now.*

After 30 seconds, say:

> *Stop now, please.*

Now reset all the pegs.

### Trial 2

Say:

> *Next you should do it with your* [nondominant] *hand, beginning from here* [upper right] *and continuing from right to left.*

If the client is left dominant, the directions are reversed. Again, show the correct S-shaped course.

> *OK, now you can begin.*

Reset all the pegs.

### Trial 3

Say:

> *Now, once again with the* [dominant] *hand. You can begin.*

Reset the pegs.

### Trial 4

Say:

*And once again with your* [nondominant] *hand. You can begin.*

## SCORING

The time limit for each trial is 30 seconds. After each trial, enter the number of correctly turned pegs on the record form (below). Record the sum of the pegs turned with the dominant and the nondominant hand in two trials (as well as dropped or upset pegs). Do not include pegs that are not placed back in the hole or that are only turned part way.

## SANTA ANA RECORD FORM

Dominant hand
  Trials 1 & 3 total | | |
  Trials 1 & 3 dropped or upset | | |
Nondominant hand
  Trials 2 & 4 total | | |
  Trials 2 & 4 dropped or upset | | |

Appendix 6

# Administration Instructions for Pursuit Aiming

The client's task is to place one dot inside each circle on the test sheet as quickly as possible. Give the client a newly sharpened pencil. The circles are small, and the test cannot be scored if the pencil is blunt. To ensure that the marks made by the client will be visible, use only B or HB pencils.

## EXTENDED PRACTICE

Make an additional copy of the test proper, using the first row for demonstration purposes, rows 2 and 3 (60 dots) for practice, and the third row (30 dots) for speeded practice. The practice material given on the front of the test blank should not be used; this differs in format from the test proper, and there are not enough items for adequate practice.

Write "Practice" at the top of the sheet, hand it to the client and say:

> *Look at this sheet with the small circles. Your job is to put a dot inside each circle. You must follow the pattern shown by the arrows.*

Indicate the pattern to the client. Say:

> *You must work as quickly as you can, but don't let your pencil touch the line of the circles. Let me show you how to do it.*

Examiner takes pencil and places a dot correctly in the first 3 or 4 circles of the top row of the practice sheet.

*You see, your dots need not be heavy, but make sure they can be easily seen. Remember not to let your pencil touch the line of the circles. Watch again.*

Examiner demonstrates with quick rapid dotting.

*Now you practice for a while placing dots here.*

Examiner indicates second and third rows, encouraging the client to work rapidly but centering each dot in the circle. When the 60 practice items are completed, give the speeded practice trial. Say:

*Good. Now do the next row as quickly and neatly as you can, then we'll be ready to do the test. Go for it!*

**Trial 1**

When the speeded practice (30 items) has been completed, write "Test" on a clean sheet and hand it to the client. Say:

*Now you are ready to start the test. It lasts for a minute, so see how many you can do. Then we'll take a rest and you'll do it again for another minute so that you can do even better after you've got used to the test.*

*Begin here* [indicate first dot], *and continue until I ask you to stop. Go as fast as you can. Please begin now.*

Discontinue after 60 seconds. Say:

*Stop now, thank you. We'll try that once again, but first let's take a little break so that you can relax your hand.*

**Trial 2**

After 30 seconds, say:

*Are you ready to start again? Please take your pencil and begin with Trial 2, here.*

The examiner points to the beginning of a row 3 or 4 rows down from the end of Trial 1.

*Begin now.*

After 60 seconds, say:

*Stop now, thank you.*

## Scoring

1. Count the sum of correct dots from both trials.
2. Count the sum of incorrect dots from both trials (those outside the circle or touching the line of the circle).
3. Count the sum of dots attempted (from the two trials combined).

Appendix 7

# Administration Instructions for Simple Reaction Time

As recommended by the WHO–NCTB *Operational Manual*, the portable battery/mains-powered Terry Reaction Time Device—which stores and displays reaction times, range, number of responses, and standard deviation—is used.

Press the reset button on the SRT device (this must be facing the examiner). Say:

> *Look at this box. You'll see a red light come on down here at the bottom. Whenever this light comes on, your job is to switch it off as quickly as you can, like this* [the examiner demonstrates].

## PRACTICE

> *Now you try. Keep this* finger [indicate the index finger of the client's dominant hand] *just touching this yellow button down here. Let's try that a few times.*

Ten trials are given, with the examiner coaching for speed throughout. As required by the WHO–NCTB *Operational Manual*, stimuli are presented at randomly varied intervals of 1 to 10 seconds. Say:

> *You see, it's like a cowboy movie. That light over there wants to shoot you. You have to kill it so quickly it doesn't get a chance. It's like a fight between you and the light. You have to shoot it before it shoots you. Sometimes the light comes on very quickly,*

*after just 1 second, and sometimes to mix you up it only comes on after 5 or even 10 seconds. So you have to be ready for it all the time.*

Switch off power.

*Now let's try some more* [switch on power again] *to see how quickly you can knock out that light.*

A further 2 minutes of practice are given, with the examiner coaching for speed for the first minute and then, as for speeded practice, falling silent.

Switch off power.

## TEST PROPER

*OK, you're doing fine. Let's do some more now* [switch on power again]. *You go as fast as you can get. This game will last for 6 minutes, so let's see if you can kill that light really fast all the time. Try to go faster and faster, but don't press till the light comes on.*

*Start now.*

The 64 trials of the test proper are now given.

After 6 minutes, the machine will shut down automatically. However, as malfunctions may occur, the examiner should set a stopwatch running as the "Start now" instruction is given and terminate the test at 7 minutes if this has not already happened. Say:

*Stop now, thank you.*

## SCORING

The SRT device will display the following scores by pressing the "roll register" button on the back of the device for each data point:

1. Number of successful trials (< 3.001 sec)
2. Number of unsuccessful trials (> 3.001 sec)
3. Mean response time of the successful trial group
4. Standard deviation of the successful response trials
5. Fastest response time (< 3.001 sec)
6. Slowest response time (< 3.001 sec)

*Appendix* 8

# Administration Instructions for the Benton Visual Retention Test (Recognition Form)

The test consists of 20 cards presented as 10 pairs of two. The first of each pair contains the pattern to be memorized and the second contains four patterns, one of which is identical to the pattern presented previously. After looking at each card for 10 seconds, the client must recognize the right patterns among the confounders on the next card presented immediately after.

## PRACTICE

Despite its apparent simplicity, the task has high novelty for non-Western clients, so that additional practice is essential. Two practice items are shown in Fig. 10.4: 10.4.1 is the first target stimulus, and 10.4.2 shows three distractors with the target item; the cards for the second item are 10.4.3 and 10.4.4, respectively.

Say:

> *I'm going to show you some cards, one at a time. Each card has one or two drawings on it, like this* (Show Item 10.4.1 to the client). *Then, I'll show you another card with four drawings on it. You have to tell me which one I showed you before.*
>
> *Let's try that now. Take another good look at this card* (10.4.1). *Tell me when you're ready* (note that this trial is not timed). *Right, now I'm taking it away and showing you this card* (place 10.4.2 on the table). *Which of these four drawings is the one you saw before?*

If the client gives the right answer, say:

*Yes, that's the right one. Now let's see why—is this part over here the same as this one here?* (Indicate a segment of 10.4.1.) *Yes, it's exacly the same. But suppose someone said that this was the right one* (Indicate one of the distractors on 10.4.2). *Why would that be wrong?*

Examiner coaches client to show where the target stimulus differs from the distractor.

If the client gives the wrong answer, say:

*No, that's not right. Look at this part over here* (indicate a segment of 10.4.1). *You see, it's not exactly the same as this one over here* (show the discrepant part of the distractor chosen by the client on 10.4.2). *You have to choose one that matches exactly. Which one do you think that is?*

Continue with this method of comparisons until the client picks the right distractor and can explain why it is correct.

*Good, I think you've got it now, so lets try another one. In the test, you get only 10 seconds to look at each drawing, so I'm going to show you this next one for 10 seconds also. Remember to look at the drawing all the time I give you, even if you think it's going to be very easy to remember.*

Again, the expectation that errors will be self-corrected must be made explicit. Say:

*Remember what I told you before: when the test begins, I'm not allowed to help you. If you give the right answer, I'll go on to the next one, but if you get it wrong—or if you don't give an answer at all—I'll wait until the time is up and then go on with the next item in the test. So please remember that if I say nothing, you should think again about your answer.*

*Are you ready? You have 10 seconds to remember the drawing—and this one's going to be more difficult because it has two drawings on it. Here it is. Remember to look at the drawing for all the time I give you.*

Place Item 10.4.3 on the table for 10 seconds exactly.

*Now—which one was it of these four* (10.4.4)*?*

Examiner coaches as before for both right and wrong answers, comparing segments and then showing the client how to eliminate wrong choices.

## TEST PROPER

*OK, I think we're ready to do the test now. Remember to look at the patterns all the time I give you. We'll start off with simple patterns, but the others will be more complicated. Ready? Here's the first one.*

Open the booklet to the first card and start the stopwatch. After 10 seconds, turn the page to the next card with the four drawings.

*Which one is it? Think carefully before you answer.*

Immediately record the client's answer [A, B, C, or D] on the Record Form. If the client does not respond within 10 to 20 seconds, say:

*Please make a choice even if you are not sure.*

If they still do not respond or say that they do not remember, mark it wrong and go on to the next card.

Show the second card to the client and proceed in the same way. Before moving to the third card, say:

*Remember to look at the patterns for the whole time I give you.*

## SCORING

The score is the number of patterns correctly recognized.

Correct answers are:

| | |
|---|---|
| 1 | D |
| 2 | A |
| 3 | C |
| 4 | C |
| 5 | B |
| 6 | D |
| 7 | B |
| 8 | A |
| 9 | A |
| 10 | C |

*Appendix* 9

# Administration Instructions for Geometric Design Reproduction

Place a copy of Fig. 10.5 on client's left (on the right for left-handed persons). Hand client some sheets of plain, unlined paper. Say:

> *Here are some different shapes. I want you to draw them for me. Make them about the same shape and size. When you draw the shapes, I want you to do it without lifting your pencil up from the paper, like this.*

Demonstrate by tracing continuously round the outline of the Greek cross without lifting your pen.

> *Please begin.*

*Appendix* **10**

# Administration Instructions for Four Words in Sequence

*I am going to say four words to you. Listen carefully, and then say them back to me in the same order.*

Say the four words at the rate of one per second.

*House—Cloud—Tree—Flying*

Record the sequence in which the words are repeated by writing 1, 2, 3, and 4 next to each word on the Record Form. If initial acquisition is imperfect, say, as appropriate:

*Good, that was two of the four,* or *Good, that was all four but not in the correct order,* etc.

Continue:

*Now I'll repeat them again in the right order and let's see if you can get them this time.*

Continue until the criterion of all four in the correct sequence is reached or five trials have been administered.

Note the clock time at which the acquisition is given, and thereafter at the top of each column the clock time for the delayed recalls. These do not have to be at the exact time stipulated, but as close as possible.

Say:

*Those four words I told you—tell them to me again.*

All four words are repeated if at 5 minutes the client remembers two or less of the original words. The administration is as follows:

*That's fine. You've got* [1 or 2] *of the words correct. I am going to repeat all four to you now. Please remember them because I will ask you again later. Do not say the words aloud after I have repeated them.*

The examiner then repeats the four words in sequence, making sure that the client does not get an extra rehearsal by repeating them. If clients attempt to repeat the words, stop them at once.

Four-Words-in-Sequence
Record Form

| | Acquisition | | | | | | 5 mins | Repeated | 10 mins | 30 mins | End of Session |
|---|---|---|---|---|---|---|---|---|---|---|---|
| | 1 | 2 | 3 | 4 | 5 | 6 | | Yes or No | | | |
| Time | | | | | | | | | | | |
| House | | | | | | | | | | | |
| Tree | | | | | | | | | | | |
| Cloud | | | | | | | | | | | |
| Flying | | | | | | | | | | | |
| | | | | | | | | | | | |
| Form B | | | | | | | | | | | |
| Road | | | | | | | | | | | |
| Wind | | | | | | | | | | | |
| Shoe | | | | | | | | | | | |
| Calm | | | | | | | | | | | |

*Appendix* **11**

# Administration Instructions for Echopraxis

## INSTRUCTIONS

*When I do this*

Tap once on the table with your flat hand

*I want you to do this*

Tap twice

*When I do this*

Tap twice

*I want you to do this.*

Tap once.

Do not say, "When I tap once, you must tap twice." This reduces the diagnostic utility of the procedure by giving the client a verbal control schema.

Check that the client has understood your instructions by alternating single and double taps a few times. If necessary, repeat the instructions and demonstration.

### Phase 1

Alternating sets: Tap once 8 or 10 times, then, without prior warning, tap twice 8 or 10 times. Make a mental note of wrong responses, but show no reaction and continue the test without pause.

### Phase 2

Following the client: Without prior warning, the examiner starts imitating whatever the client does. Thus, you follow the client's single tap with a single, and a double with a double. However, if echopraxia is present, this method will not work, and you will then have to imitate what the client ought to have done. Continue Phase 2 for about 60 seconds.

## SCORING

5. Quick and error free
4. Hesitant but error free
3. Occasional errors which are self-corrected
2. Up to half errors, most uncorrected
1. Wrong most or all of the time
9. Cannot be tested

*Appendix* 12

# Administration Instructions for the Tower of London

## 1. TOWER OF LONDON

Point to the apparatus and say:

*Here we have three pegs, each a different length, and three beads in different colours. The beads can be arranged on the pegs to make patterns.*

Take the beads off the pegs and put them down on the table. Then place the "Start Position" card between the client and the apparatus.

*The picture on this card shows one pattern. Please copy it with the beads?*

Remove the "Start Position" card, keeping it next to you for your own reference, and place the "Practice Item 1" card between the client and the apparatus.

*Now I'm going to show you another pattern. I'll ask you to change the beads from the pattern here* (point to the apparatus) *to this new pattern* (point to card). *The rules are very easy:*

* *You can only remove one bead at a time*

* *You can only move from one peg straight onto another peg. You can't put the beads down on the table, and you can't have more than one bead in your hands at the same time.*

* *You can only put one bead on this peg, two beads on this peg, but of course all three beads could go onto this peg* (point in turn to Pegs 1, 2, and 3).

Demonstrate this by pretending to put two beads onto Peg 1:

*You see, this top one wouldn't really be on the peg at all, it would just fall off.*

*Each problem must be solved in a certain number of moves. When I give you the card I'll tell you how many moves are allowed. If you don't get it out in the correct number of moves, you can always go back to the starting position and try again.*

*Now try this first example for me. You see you are allowed three moves for this one.*

If your client solves this first problem quickly and easily, then go straight on to Practice Item 3, saying:

*Very good, now try another example. See if you can get this one out in three moves.*

However, if clients have problems, help them to do the practice item, and then try again on their own. In such cases, administer Practice Item 2 also.

On all the practice cards, coach as much as necessary.

The standard starting position, the practice items, and the test items are shown here. Before administering the test, you must make up a set of 17 cards on which colored circles have been cut out of red, green, and blue paper and glued to the stimulus cards over a line drawing of the apparatus, as in Fig. A12.1. Note that on the position cards, you substitute colored circles for the letters R, G, and B.

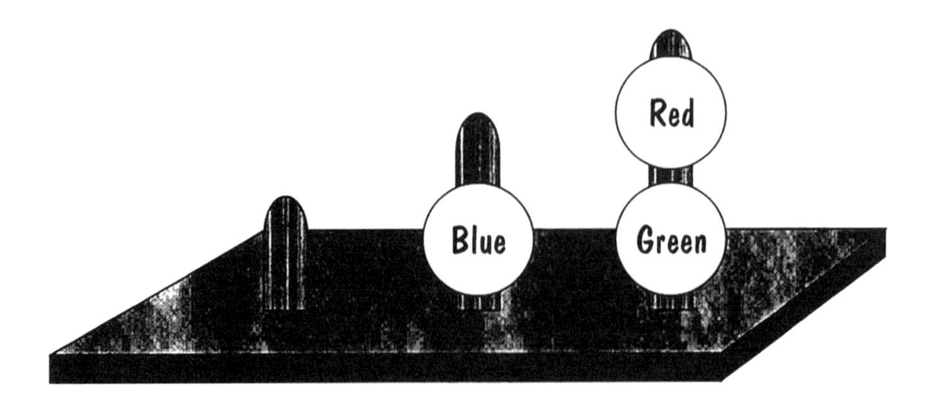


FIG. A12.1. Specimen position card for the Tower of London Test showing starting position.

| | | | |
|---|---|---|---|
| *Starting position:* | 1 – | 2 B | 3 R G |
| *Practice items:* | | | |
| Practice Item 1 (2 Moves) | 1 B | 2 G | 3 R |
| Practice Item 2 (3 Moves) | 1 R | 2 B | 3 G |
| Practice Item 3 (3 Moves) | 1 G | 2 B | 3 R |
| Practice Item 4 (5 Moves) | 1 – | 2 G | 3 B R |
| *Test items:* | | | |
| 1. (2 moves) | 1 R | 2 GB | 3 – |
| 2. (2 moves) | 1 B | 2 R | 3 G |
| 3. (3 moves) | 1 – | 2 R | 3 BG |
| 4. (3 moves) | 1 – | 2 BR | 3 G |
| 5. (4 moves) | 1 – | 2 B | 3 GR |
| 6. (4 moves) | 1 – | 2 GR | 3 B |
| 7. (4 moves) | 1 G | 2 – | 3 BR |
| 8. (4 moves) | 1 G | 2 BR | 3 – |
| 9. (5 moves) | 1 – | 2 – | 3 BGR |
| 10. (5 moves) | 1 – | 2 – | 3 GBR |
| 11. (5 moves) | 1 – | 2 R | 3 GB |
| 12. (5 moves) | 1 G | 2 R | 3 B |

## 2. NEW TOWER OF LONDON

This test is not in the public domain. See Appendix 4 for ordering information.

*Appendix* **13**

# Administration Instructions for Writing to Dictation

## TELEPHONE NUMBERS TO DICTATION

Read the numbers at the rate you normally use for leaving your own telephone number on an answering machine or with a telephonist. Repetition is not allowed. If the client asks you to repeat a number, say, "Just do the best you can."

| Target Number | Client's Response | Score (0 or 1) |
|---|---|---|
| 8595 | | |
| 55 8358 | | |
| 646 4275 | | |
| 418 7903 | | |
| 938 3721 | | |
| 724 6019 | | |
| 442 5390 | | |
| 726 5981 | | |
| 317 4925 | | |
| 883 7204 | | |

## WORDS AND SENTENCES

*Here is some paper for you* [hand the client a few sheets of unlined paper]. *I'm going to read you some words and then some sentences. Please write them down for me. Write in handwriting with the letters joined together, not in capitals:*

Words and phrases:

clock
square
zoo
sugar
cough
quickly
Colonel Grice
7 kg
120 km/h
1 o'clock
He shouted the warning.

Sentences:

Mr. Oosthuizen *[substitute a local last name of equivalent difficulty]* advertises his watch business in the yellow pages.
Johnny rides his bicycle to school through the mountain pass.

# References

Abikoff, H., Alvir, J. I., Hong, G., Suckoff, R., Orazio, J., Solomon, S., & Saravay, S. (1987). Logical memory subtest of the Wechsler Memory Scale: Age and education norms and alternate form reliability of two scoring systems. *Journal of Clinical and Experimental Psychology, 9*, 435–448.

Adams, R. A., & Victor, M. (1989). *Principles of neurology* (4th ed.). New York: McGraw-Hill.

Adan, M. (1997). *Traumatic brain injury in an identical twin.* Unpublished manuscript.

Adonisi, M. (1988). Paragraph recall in Black clerical workers and students. Unpublished data.

Albee, G. W. (1977). The Protestant Ethic, sex and psychotherapy. *American Psychologist, 32*, 150–161.

Albee, G. W. (1986). Toward a just society: Lessons from observations on the primary prevention of psycho-pathology. *American Psychologist, 41*, 891–898.

Anastasi, A. (1988). *Psychological testing* (6th ed.). New York: Macmillan.

Anderson, S. J., & Macpherson, A. (in draft). *Normative test data for Zulu-speakers: Selected neuropsychological measures.* Psychology Department, University of Natal, Pietermaritzburg.

Andersson, N., & Marks, S. (1989). The state, class and the allocation of health resources in southern Africa. *Social Science Medicine, 28*, 515–530.

Anger, W. K., Cassitto, M. G., Liang, Y. X., Amador, R., Hooisma, J., Chrislip, D. W., Mergler, D., Keifer, M., Hortnagl, J., & Fournier, L. (1993). Comparison of performance from three continents on the WHO-Recommended Neurobehavioral Core Test Battery. *Environmental Research, 62*, 125–147.

Annegers, J. F., Grabow, J. D., Kurland, L. T., & Laws, E. R. (1980). The incidence, causes, and secular trends of head trauma in Olmsted County, Minnesota, 1935–1974. *Neurology, 30*, 912–919.

Ardila, A., Roselli, M., & Puente, A. E. (1994). *Neuropsychological evaluation of the Spanish speaker.* New York: Plenum.

Artiola i Fortuny, L. (1996). Review of Neuropsychological evaluation of the Spanish-speaker. *The Clinical Neuropsychologist, 10*, 229–234.

Avenant, T. J. (1988). *The establishment of an individual intelligence scale for adult South Africans. Report on an exploratory study conducted with WAIS–R on a sample of Blacks* (Report No. P-91). Pretoria: Human Sciences Research Council.

Bach-y-Rita, G., Lion, J. R., Climent, C. E., & Erin, F. R. (1971). Episodic dyscontrol: A study of 130 violent patients. *American Journal of Psychiatry, 127,* 1473–1478.

Baddeley, A. D. (1986). *Working memory.* Oxford, England: Clarendon.

Barnes, B., & Bloor, D. (1982). Relativism, rationalism and the sociology of knowledge. In M. Hollis & S. Lukes (Eds.), *Rationality and relativism* (pp. 21–47). Oxford, England: Basil Blackwell.

Barrett, G. V., & Depinet, R. L. (1991). A reconsideration of testing for competence rather than for intelligence. *American Psychologist, 46,* 1012–1024.

Beatty, W. W., & Scott, J. G. (1993). Issues and developments in the neuropsychological assessment of patients with multiple sclerosis. *Journal of Neurologic Rehabilitation, 7,* 87–97.

Bennet, J., & George, S. (1987). *The hunger machine.* Cambridge, England: Polity Press.

Benton, A. L. (1989). Historical notes on the concussion syndrome. In H. S. Levin, H. M. Eisenburg, & A. L. Benton (Eds.), *Mild head injury* (pp. 3–7). New York: Oxford University Press.

Berg, R., Franzen, M., & Wedding, D. (1987). *Screening for brain impairment: A manual for mental health practice.* New York: Springer.

Berlyne, D. E. (1974). *Studies in the new experimental aesthetics: Steps toward an objective psychology of aesthetic appreciation.* Washington, DC: Hemisphere.

Berry, J. W. (1966). Temne and Eskimo perceptual skills. *International Journal of Psychology, 1,* 207–229.

Berry, J. W. (1972). Radical cultural relativism and the concept of intelligence. In L. J. Cronbach & P. J. D. Drenth (Eds.), *Mental tests and cultural adaptation* (pp. 77–88). The Hague, Netherlands: Mouton.

Berry, J. W. (1976). *Human ecology and cognitive style.* New York: Wiley.

Berry, J. W. (1984). Towards a universal psychology of cognitive competence. *International Journal of Psychology, 19,* 335–361.

Berry, J. W. (1988). Cognitive and social factors in psychological adaptation to acculturation among the James Bay Cree. In G. K. Verma & C. Bagley (Eds.), *Cross cultural studies of personality and cognition* (pp. 111–142). London: Macmillan.

Berry, J. W. (1989). Imposed emics-derived etics: The operationalisation of a compelling idea. *International Journal of Psychology, 24,* 721–735.

Biesheuvel, S. (1972). Adaptability: Its measurement and determinants. In L. J. Cronbach & P. J. D. Drenth (Eds.), *Mental tests and cultural adaptation* (pp. 47–62). The Hague, Netherlands: Mouton.

Biesheuvel, S., & Liddicoat, R. (1959). The effects of cultural factors on intelligence-test performance. *Journal of the National Institute for Personnel Research, 8,* 3–14.

Blakeley, T. A., & Harrington, D. E. (1993). Mild head injury is not always mild: Implications for damage litigation. *Medical Science and the Law, 33,* 231–242.

Bloom, H. (1994). *The Western canon: The books and school of the ages.* New York: Harcourt Brace.

Blumbergs, P. C., Scott, G., & Manavis, J. (1995). Topography of axonal injury as defined by amyloid precursor protein in the sector scoring method in mild and severe head injury. *Journal of Neurotrauma, 12,* 565–572.

Boas, F. (1911). *Handbook of American Indian languages.* Washington, DC: U.S. Government Printing Office.

Boeyens, J. (1989). *Learning potential: A theoretical perspective* (Report No. Pers-432). Pretoria: Human Sciences Research Council.

Bornstein, R. A. (1985). Normative data on selected neuropsychological measures from a nonclinical sample. *Journal of Clinical Psychology, 41,* 651–659.

Bornstein, R. A. (1986a). Classification rates obtained with "standard" cut-off scores on selected neuropsychological measures. *Journal of Clinical and Experimental Neuropsychology, 8,* 413–420.

Bornstein, R. A. (1986b). Normative data on intermanual differences on three tests of motor performance. *Journal of Clinical and Experimental Neuropsychololgy, 8*, 12–20.

Bouchard, T. J., Lykken, D. T., McGue, M., Segal, N. L., & Tellegen, A. (1990). Sources of human psychological differences: The Minnesota study of twins reared apart. *Science, 250*, 223–228.

Brody, N. (1992). *Intelligence* (2nd ed.). San Diego: Academic Press.

Brown, D. S. O. (1991). HIV infection in persons with prior mental retardation. *AIDS Care, 3*, 165–173.

Brown, D. S. O., & Nell, V. (1991). The epidemiology of traumatic brain injury in Johannesburg: I. Methodological issues in a developing country context. *Social Science and Medicine, 33*, 283–287.

Brown, D. S. O., Wills, C. E., Yousefi, V., & Nell, V. (1991). Neurotoxic effects of chronic exposure to manganese dust. *Neuropsychiatry, Neuropsychology and Behavioural Neurology, 4*, 238–248.

Bulhan, H. (1985). *Frantz Fanon and the psychology of oppression.* New York: Plenum.

Butters, N., Grant, I., Haxby, J., & Judd, L. L. (1990). Assessment of AIDS-related cognitive changes: Recommendations of the NIMH Workshop on Neuropsychological Assessment Approaches. *Journal of Clinical and Experimental Neuropsychology, 12*, 963–978.

Carroll, J. B. (1983). Studying individual differences in cognitive abilities: Implications for cross-cultural studies. In S. H. Irvine & J. W. Berry (Eds.), *Human assessment and cultural factors* (pp. 213–236). New York: Plenum.

Carroll, J. B. (1984). Raymond B. Cattell's contribution to the theory of cognitive abilities. *Multivariate Behavioral Research, 19*, 300–306.

Carroll, J. B. (1993). *Human cognitive abilities: A survey of factor analytical studies.* New York: Cambridge University Press.

Cassito, M. G., Camerino, D., Hanninen, H., & Anger, W. K. (1990). International collaboration to evaluate the WHO neurobehavioral core test battery. In B. L. Johnson (Ed.), *Advances in neurobehavioral toxicology: Applications in environmental and occupational health.* Chelsea, MI: Lewis.

Cattell, R. B. (1967). The theory of fluid intelligence and crystallized intelligence in relation to "culture fair" tests and its verification in children 9–12 years old. *Revue de Psychologie Appliquee, 17*, 135–154.

Channabavasanna, S. M., Gururaj, G., Das, R. S., & Kaliaperumal, V. G. (1994). *Epidemiology of head injuries.* Bangalore: National Institute of Mental Health and Neurosciences.

Chia, S. E., Jeyaratnam, J., Ong, C. N., Ng., T. P., & Lee, H. S. (1994). Impairment of color vision among workers exposed to low concentrations of styrene. *American Journal of Industrial Medicine, 26*, 481–488.

Chomsky, N. (1959/1964). A review of B. F. Skinner's *Verbal Behavior.* In J. A. Fodor & J. J. Katz (Eds.), *The structure of language: Readings in the philosophy of language.* Englewood Cliffs, NJ: Prentice-Hall.

Chouinard, M. J., & Rouleau, I. (1997). The 48-pictures test: A two-alternative forced-choice recognition test for the detection of malingering. *Journal of the International Neuropsychological Society, 3*(6), 545–552.

Christensen, A. L. (1974). *Luria's neuropsychological investigation.* Copenhagen: Munksgaard.

Cicerone, K. D., & Tupper, D. E. (1986). Cognitive assessment in the neuropsychological rehabilitation of head injured adults. In B. Uzzell & Y. Gross (Eds.), *Clinical neuropsychology of intervention.* Boston: Nijhoff Hingham.

Claassen, N. C. W. (1985). *Socio-economic Deprivation Questionnaire* (Document 2387). Pretoria: Human Sciences Research Council.

Claassen, N. C. W. (1997). Cultural differences, politics and test bias in South Africa. *European Review of Applied Psychology, 47*, 297–307.

Claxton, G. (Ed.). (1988). *Growth points in cognition.* London: Routledge.

Cochrane, H. J., Baker, G. S., & Meudell, P. R. (1998). Simulating a memory impairment: Can amnesics implicitly outperform simulators? *British Journal of Clinical Psychology, 37,* 31–48.

Cole, M., & Scribner, S. (1974). *Culture and thought, a psychological introduction.* New York: Wiley.

Colvin, M., Myers, J., Nell, V., Rees, D., & Cronje, R. (1994). A cross-sectional survey of neurobehavioural effects of chronic solvent exposure in a paint manufacturing plant. In S. Araki (Ed.), *Neurobehavioral methods and effects in occupational and environmental health* (pp. 181–192). San Diego: Academic Press.

Conrad, J. (1902/1960). *Heart of darkness.* New York: Bantam.

Cosmides, L., Tooby, J., & Barkow, J. H. (1992). Evolutionary psychology and conceptual integration. In J. H. Barkow, L. Cosmides, & J. Tooby (Eds.), *The adapted mind: Evolutionary psychology and the generation of culture.* New York: Oxford University Press.

Crawford-Nutt, D. H. (1976). Black scores on Raven's standard progressive matrices: An artifact of method of test presentation. *Psychologia Africana, 16,* 201–206.

Crawford-Nutt, D. H. (1977a). Assessing the intellectual capacity of subjects in cultural transition. In J. H. Poortinga (Ed.), *Basic problems in cross-cultural psychology* (pp. 49–59). Amsterdam & Lisse: Swets & Zeitinger.

Crawford-Nutt, D. H. (1977b). The effect of educational level on the test scores of people in South Africa. *Psychologia Africana, 17,* 49–59.

Crawford-Nutt, D. H. (1977c). *The Symco Test: Research report and guide to its administration and scoring* (Report No. Pers-165). Johannesburg: National Institute for Personnel Research Council for Scientific and Industrial Research.

Cronbach, L. J., & Meehl, P. E. (1955). Cultural validity in psychological tests. *Psychological Bulletin, 52,* 281–302.

Dabydeen, D. (Ed.). (1985). *The Black presence in English literature.* Manchester, England: Manchester University Press.

Dague, P. (1972). Development, application and interpretation of tests for use in French-speaking black Africa and Madagascar. In L. J. Cronbach & P. J. D. Drenth (Eds.), *Mental tests and cultural adaptation* (pp. 63–74). The Hague: Mouton.

Damassio, A. R. (1994). *Descartes' error: Emotion, reason, and the human brain.* New York: Avon.

Darwin, C. (1859/1902). *On the origin of species by means of natural selection, or the preservation of favoured races in the struggle for life.* London: Grant Richards.

Dasen, P. R. (1977). Cross-cultural cognitive development: The cultural aspects of Piaget's theory. *Annals of the New York Academy of Sciences, 285,* 332–337.

Dasen, P. R. (1981). "Strong" and "Weak" universals: Sensori-motor intelligences and concrete operations. In B. Lloyd & J. Gay (Eds.), *Universals of human thought* (pp. 137–156). Cambridge, England: Cambridge University Press.

Dasen, P. R. (1984). The cross-cultural study of intelligence: Piaget and the Baoulé. *International Journal of Psychology, 19,* 407–434.

Dasen, P. R., Lavallée, M., & Retschitzki, J. (1979). Training conservation of quantity (liquids) in West African (Baoulé) children. *International Journal of Psychology, 14,* 576–68.

Delis, D. C., Kramer, J. H., Fridlund, A. J., & Kaplan, E. (1990). A cognitive science approach to neuropsychological assessment. In P. McReynolds & J. C. Rosen (Eds.), *Advances in psychological assessment* (Vol. 7, pp. 101–132). New York: Plenum.

Dorfman, D. D. (1982). Henry Goddard and the feeble-mindedness of Jews, Hungarians, Italians, and Russians. *American Psychologist,* 37, 96–97.

Ekstrom, R. B., French, J. W., & Harman, H. H. (1979). Cognitive factors: Their identification and replication. *Multivariate Behavioral Research Monographs, 79.*

Elliott, F. A. (1982). The dyscontrol syndrome. *Journal of Nervous and Mental Diseases, 170,* 680–687.

Escalona, E., Yanes, L., Feo, O., & Maizlish, N. (1995). Neurobehavioral evaluation of Venezuelan workers exposed to organic solvent mixtures. *American Journal of Industrial Medicine, 27*, 15–27.

Fanon, F. (1968). *Wretched of the earth*. New York: Grove.

Fasteneau, P. S., & Adams, K. M. (1996). Heaton, Grant and Matthews' comprehensive norms: An overzealous attempt. *Journal of Clinical and Experimental Neuropsychology, 18*, 444–448.

Feuerstein, R. (1979). *The dynamic assessment of retarded performers*. Baltimore: University Park Press.

Feuerstein, R. (1980). *Instrumental enrichment: An intervention program for cognitive modifiability*. Baltimore: University Park Press.

Feuerstein, R., Rand, Y., Hoffman, M., & Miller, R. (1979). Cognitive modifiability in retarded adolescents: Effects of instrumental enrichment. *American Journal of Mental Deficiency, 83*, 539–550.

Finlayson, M. A. J., Johnson, K. A., & Reitan, R. M. (1977). Relationship of level of education to neuropsychological measures in brain-damaged and non-damaged adults. *Journal of Consulting and Clinical Psychology, 35*, 536–542.

Flavell, J. H. (1963). *The developmental psychology of Jean Piaget*. Princeton, NJ: Van Nostrand.

Fleishman, E. A. (1954). Dimensional analysis of psychomotor abilities. *Journal of Experimental Psychology, 48*, 437–454.

Foucault, M. (1963/1989). *The birth of the clinic*. London: Routledge.

Foucault, M. (1980). *Power/knowledge: Selected interviews and other writings, 1972–1977*. London: Harvester.

Frank, G. (1983). *The Wechsler enterprise: An assessment of the development, structure and use of the Wechsler tests of intelligence*. New York: Pergamon.

Freeman, M. C. (1984). *The effect of cultural variables on the Goodenough Harris drawing test and the Standard Progressive Matrices*. Unpublished master's thesis, University of the Witwatersrand, Johannesburg, South Africa.

French, J. W. (1951). *The description of aptitude and achievement tests in terms of rotated factors*. Chicago: University of Chicago Press.

Fuld, P. A. (1977). *Fuld object-memory evaluation*. Chicago: Stoelting.

Gardner, H. (1987). *The mind's new science*. New York: Basic Books.

Geertz, C. (1971). *The interpretation of cultures*. New York: Basic Books.

Geffen, G. M., O'Hanlon, K. J., Clark, C. R. et al. (1990). Performance measures of 16 to 86-year-old males and females on the Auditory Verbal Learning Test. *Clinical Neuropsychologist, 4*, 45–63.

Gellatly, A., Rogers, D., & Sloboda, J. A. (Eds.). (1989). *Cognition and social worlds*. Oxford: Clarendon Press.

Gellner, E. (1981). Relativism and universals. In B. Lloyd & J. Gay (Eds.), *Universals of human thought* (pp. 1–20). Cambridge, England: Cambridge University Press.

Gilbert, A. J. (1986). *Psychology and social change in the third world: A cognitive perspective*. Unpublished doctoral dissertation, University of South Africa, Pretoria.

Gilbert, A. J. (1989). Things fall apart? Psychological theory in the context of rapid social change. *S.A. Journal of Psychology, 19*, 91–100.

Goldstein, K. (1942). *After effects of brain injuries in war*. New York: Grane & Stratton.

Gould, S. J. (1981). *The mismeasure of man*. Harmondsworth: Penguin.

Grant, G. V. (1970). Spatial thinking: A dimension in African intellect. *Psychologia Africana, 13*, 222–239.

Grant, G. V. (1972). Conceptual reasoning: Another dimension of African intellect. *Psychologia Africana, 14*, 170–185.

Griffin, G. A. E., Glassmire, D. M., Henderson, E. A., & McCann, C. (1997). Rey II: Redesigning the Rey screening test of malingering. *Journal of Clinical Psychology, 53*(7), 757–766.

Grimm, R. J., Hemenway, W. G., Lebray, P. R., & Black, F. O. (1989). The perilymph fistula syndrome defined in mild head trauma. *Acta Otolaryngol. Supplement. Stockholm, 464*, 1–40.

Guilford, J. P. (1956). The structure of intellect. *Psychological Bulletin, 53*, 267–293.

Hall, E. (1989). *Inventing the barbarian: Greek self-definition through tragedy.* Oxford, England: Clarendon.

Halstead, W. C. (1947). *Brain and intelligence: A quantitative study of the frontal lobes.* Chicago: University of Chicago Press.

Hanes, K. R. (1996). *New Tower of London test manual.* Carlton, Victoria, Autralia: ACCTA.

Heaton, R. K., Grant, I., & Matthews, C. G. (1991). *Comprehensive norms for an expanded Halstead–Reitan Battery: Demographic corrections, Research findings, and clinical applications.* Odessa, FL: Psychological Assessment Resources.

Herrnstein, R. J., & Murray, C. (1994). *The bell curve: Intelligence and class structure in American life.* New York: The Free Press.

Hick, W. (1952). On the rate of gain of information. *Quarterly Journal of Experimental Psychology, 86*, 10–13.

Hofmeyer, I. (1991). Popularising history: The case of Gustav Preller. In R. Hill, M. Miller, & M. Trump (Eds.), *African Studies Forum* (pp. 49–76). Pretoria: Human Sciences Research Council.

Hollis, M., & Lukes, S. (Eds.). (1982). *Rationality and relativism.* Oxford, England: Basil Blackwell.

Holtzman, W. H., Evans, R. I., Kennedy, S. W., & Iscoe, I. (1987). Psychology and health: Contributions of psychology to the improvement of health and health care. *International Journal of Psychology, 22*, 221–267.

Hudson, W. (1960). "Pictorial depth perception in subcultural groups in Africa." *Journal of Social Psychology, 52*, 183–208.

Hunt, E. (1995). The role of intelligence in modern society. *American Scientist, 83*, 356–368.

Inhelder, B., & Piaget, J. (1958). *The growth of logical thinking from childhood to adolescence.* Paris: Basic Books.

Irvine, S. H. (1966). Towards a rationale for testing attainments and abilities in Africa. *British Journal of Educational Psychology, 36*, 24–32.

Irvine, S. H. (1969). Figural tests of reasoning in Africa: Studies in the use of the Raven's Progressive Matrices across cultures. *International Journal of Psychology, 4*, 217–228.

Irvine, S. H., & Berry, J. W. (Eds.). (1988a). *Human abilities in cultural context.* Cambridge, England: Cambridge University Press.

Irvine, S. H., & Berry, J. W. (1988b). The abilities of mankind: A revaluation. In S. H. Irvine & J. W. Berry (Eds.), *Human abilities in cultural context* (pp. 3–59). Cambridge, England: Cambridge Univeristy Press.

Ivinskis, A., Allen, S., & Shaw, E. (1971). An extension of the Wechsler Memory Scales to lower age groups. *Journal of Clinical Psychology, 27*, 354–357.

Jahoda, G. (1981). Pictorial perception and the problem of universals. In B. Lloyd & J. Gay (Eds.), *Universals of human thought* (pp. 25–45). Cambridge, England: Cambridge University Press.

Jansen, C. P. (1988). *Psychosocial consequences after brain trauma. A follow up study of individuals and their families.* Unpublished master's thesis, University of South Africa.

Jennett, B. (1989). Some international comparisons. In H. S. Levin, H. M. Eisenburg, & A. L. Benton (Eds.), *Mild head injury* (pp. 23–34). New York: Oxford University Press.

Jensen, A. R. (1969). How much can we boost IQ and scholastic achievement? *Harvard Educational Review, 39*, 1–123.

Jensen, A. R. (1980). *Bias in mental testing.* London: Methuen.

Jensen, A. R. (1984). Mental speed and levels of analysis. *The Behavioural and Brain Sciences, 7*, 295–296.

Jensen, A. R. (1988). Speed of information processing and population differences. In S. H. Irvine & J. W. Berry (Eds.), *Human abilities in cultural context* (pp. 105–145). Cambridge, England: Cambridge University Press.

Kagan, J., Klein, R. E., Finlay, G. E., & Rogoff, B. (1977). A study in cognitive development. *Annals New York Academy of Sciences, 285*, 374–388.

Kahneman, D. (1973). *Attention and effort.* Englewood Cliffs, NJ: Prentice-Hall.

Kamin, L. J. (1974). *The science and politics of I.Q.* Potomac, MD: Lawrence Erlbaum Associates.

Kaplan, E. (1988). A process approach to neuropsychological assessment. In T. Boll & B. K. Bryant (Eds.), *Clinical neuropsychology and brain function: Research, measurement, and practice.* Washington, DC: American Psychological Association.

Kendall, I. M., Verster, M. A., & von Mollendorf, J. W. (1988). Test performance of Blacks in Southern Africa. In S. H. Irvine & J. W. Berry (Eds.), *Human abilities in cultural context* (pp. 239–299). Cambridge, England: Cambridge University Press.

Knox, J. E. (1993). Translator's introduction. In L. Vygotsky & A. R. Luria (Eds.), *Studies on the history of behaviour: Ape, primitive and child* (pp. 1–35). Hillsdale NJ: Lawrence Erlbaum Associates.

Kozulin, A. (1984). *Psychology in utopia.* Cambridge, MA: MIT Press.

Laboratory of Comparative Human Cognition (1982). Culture and intelligence. In R. J. Sternberg (Ed.), *Handbook of human intelligence* (pp. 642–719). Cambridge, England: Cambridge University Press.

Langenhoven, H. P. (1960). Comments on "The effects of cultural performance on intelligence test performance." *Journal of the National Institute for Personnel Research, 8*, 150–152.

Lee, S. H., & Lee, S. H. (1993). A study on the neurobehaviourioul effects of occupational exposure to organic solvents in Korean farm workers. *Environmental Research, 60*, 227–232.

Leininger, B. E., Gramling, S. E., Farell, A. D., Kreitzer, J. S., & Peck, E. A. (1990). Neurophyschological deficits in symptomatic minor head injury patients after concussion and mild concussion. *Journal of Neurology, Neurosurgery and Psychiatry, 53*, 293–296.

Lezak, M. D. (1983). *Neuropsychological assessment* (2nd ed.). New York: Oxford University Press.

Lezak, M. D. (1995). *Neuropsychological assessment* (3rd ed.). New York: Oxford University Press.

Liang, Y., Chen, Z., Sun, R., Fang, Y., & Yu, J. (1990). Application of the WHO Neurobehavioural Core Test Battery and other neurobehavoural screening methods. In B. L. Johnson (Ed.), *Advances in neurobehavioural toxicology: Applications in environmental and occupational health.* Chelsea, MI: Lewis.

Liang, Y. X., Sun, R. K., Sun, Y., Chen, Z. Q., & Li, L. H. (1993). Psychological effects of low exposure to mercury vapor: Application of a computer-administered neurobehavioral evaluation system. *Environmental Research, 60*, 320–327.

Lloyd, B., & Gay, J. (Eds.). (1981). *Universals of human thought.* Cambridge, England: Cambridge University Press.

London, L., Myers, J. E., Nell, V., Taylor, T. R., & Thompson, M. L. (1997). An investigation into neurological and neurobehavioural effects of long-term agrichemical use among deciduous fruit farm workers in the Western Cape, South Africa. *Environmental Research, 73*, 132–145.

London, L., Nell, V., Thompson, M. L., & Myers, J. E. (1998). Health status among farm workers in the Western Cape: Collateral evidence from a study of occupational hazards. *South African Medical Journal, 88*, 1096–1101.

Lorenz, K. Z. (1963). *On aggression.* New York: Harcourt.

Luria, A. R. (1933). The second psychological expedition to Central Asia. *Science, 78*, 191–192.

Luria, A. R. (1962/1980). *Higher cortical functions in man.* New York: Basic Books.

Luria, A. R. (1963). *Restoration of function after brain injury.* New York: Macmillan.

Luria, A. R. (1965). L. S. Vygotsky and the problem of localization of functions. *Neuropscyhologia, 3,* 387–392.

Luria, A. R. (1970). *Traumatic aphasia.* The Hague: Mouton. (Original work published 1947)

Luria, A. R. (1971). Toward the problem of the historical nature of psychological processes. *International Journal of Psychology, 6,* 2260–2272.

Luria, A. R. (1973). *The working brain: An introduction to neuropsychology.* London: Penguin.

Luria, A. R. (1975a). *The man with a shattered world: The history of a brain wound.* Harmondsworth: Penguin.

Luria, A. R. (1975b). *The mind of a mnemonist.* Harmondsworth: Penguin.

Luria, A. R. (1976a). *Basic problems of neuro-linguistics.* The Hague: Mouton.

Luria, A. R. (1976b). *Cognitive development: Its cultural and social foundations.* Cambridge, MA: Harvard University Press.

Luria, A. R. (1976c). *The neuropsychology of memory.* Washington, DC: Winston.

Luria, A. R. (1979). *The making of mind.* Cambridge, MA: Harvard University Press.

Maddocks, D. L., Saling, M., & Dicker, G. D. (1995). A note on normative data on a test sensitive to concussion in Australian Rules footballers. *Australian Psychologist, 30,* 125–127.

Maizlish, N. A., Parra, G., & Feo, O. (1995). Neurobehavioral evaluation of Venezuelan workers exposed to inorganic lead. *Occupational and Environmental Medicine, 52,* 408–414.

Maj, M., Satz, P., Janssen, R., Zaudig, M., et al. (1994). WHO Neuropsychiatric AIDS Study, Cross-Sectional Phase I: Study design and psychiatric findings. *Archives of General Psychiatry, 51,* 51–61.

Makunga, N. V. (1988). *The development of provisional norms for Black South Africans on selected neuropsychological tests and their clinical validation.* Unpublished doctoral dissertation, University of South Africa.

Manganyi, N. C. (1991). *Treachery and innocence: Psychology and racial difference in South Africa.* Johannesburg: Ravan.

Mannoni, O. (1956). *Prospero and Caliban: The psychology of colonization.* London: Methuen.

Marion, D. W., & Carlier, P. M. (1994). Problems with initial Glasgow Coma Scale assessment caused by prehospital treatment of patients with head injuries: Results of the national survey. *Journal of Trauma, 36,* 89–95.

Marshall, L. F., & Marshall, S. B. (1996). *Minor head injury: Differing manifestations.* Unpublished manuscript.

Marshall, L. F., & Ruff, R. M. (1989). Neurosurgeon as victim. In H. S. Levin, H. M. Eisenburg, & A. L. Benton (Eds.), *Mild head injury* (pp. 276–280). New York: Oxford University Press.

Matarazzo, J. D. (1972). *Wechsler's measurement and appraisal of adult intelligence* (5th ed.). New York: Oxford University Press.

Matarazzo, J. D. (1990). Psychological assessment versus psychological testing: Validation from Binet to the school, clinic, and courtroom. *American Psychologist, 45,* 999–1017.

Matarazzo, J. D., & Herman, D. O. (1984). Base-rate data for the WAIS–R: Test–retest stability and VIQ–PIQ differences. *Journal of Clinical Neuropsychology, 6,* 351–346.

Matarazzo, J. D., & Herman, D. O. (1985). Clinical uses of the WAIS–R: Base-rate differences between VIQ and PIQ in the WAIS–R standardisation sample. In B. B. Wolman (Ed.), *Handbook of intelligence: Theories, measurements, and applications* (pp. 899–932). New York: Wiley.

Matarazzo, J. D., & Prifitera, A. (1989). Subtest scatter and premorbid intelligence: Lessons from the WAIS–R standardization sample. *Psychological Assessment, 1,* 186–191.

McClelland, D. C. (1961). *The achieving society.* New York: The Free Press.

McClelland, D. C. (1973). Testing for competence rather than for intelligence. *American Psychologist, 28,* 1–14.

McCrory, P. R., & Berkowic, S. F. (1998). Second impact syndrome. *Neurology, 50,* 677–683.

McLennan, G. (1981). *Marxism and the methodologies of history.* London: Verso.

McNair, D. M., Lorr, M., & Droppelman, L. F. (1971). *Profile of mood states.* San Diego, CA: Educational and Industrial Testing Services.

McNair, D. M., Lorr, M., & Droppelman, L. F. (1992). *Manual for the Profile of Mood States (POMS)–Revised.* San Diego, CA: Educational and Industrial Testing Services.

Melendez, F. (1994). The Spanish version of the WAIS: Some ethical considerations. *The Clinical Neuropsychologist, 8,* 388–393.

Mercer, J. R. (1984). What is a racially and culturally nondiscriminatory test? A sociological and pluralistic perspective. In C. R. Reynolds & R. T. Brown (Eds.), *Perspectives on "bias in mental testing".* New York: Plenum.

Miller, R. (1984). *Reflections of mind and culture.* Pietermaritzburg: University of Natal Press.

Mokhuane, E. M. Q. (1997). *Pain: Psychological measurement and treatment.* Unpublished doctoral dissertation, University of South Africa, Pretoria.

Monopolis, S., & Lion, J. R. (1983). Problems in the diagnosis of intermittent explosive disorder. *American Journal of Psychiatry, 140,* 1200–1202.

Monroe, R. R. (1975). Anticonvulsants in the treatment of aggression. *Journal of Nervous and Mental Diseases, 160,* 119–126.

Moselenyane, N. F. (1990). *The neuropsychological assessment of traumatically brain-injured adults in Lebowa and their after-care services.* Unpublished master's thesis, University of South Africa.

Moses, J. A., Golden, C. J., Ariel, R., & Gustavson, J. L. (1983). *Interpretation of the Luria–Nebraska Neuropsychological Battery* (Vol. 1). New York: Grune & Stratton.

Mundy-Castle, A. (1983). Are Western psychology concepts valid in Africa? A Nigerian review. In S. H. Irvine & J. W. Berry (Eds.), *Human assessment of cultural factors.* New York: Plenum.

Mungas, D. (1983). An empirical analysis of specific syndromes of violent behaviour. *Journal of Nervous and Mental Disease, 171,* 354–361.

Murdoch, B. D. (1982). Verbal and performance I.Q. differences in the determination of brain damage using the SAWAIS. *South African Journal of Psychology, 12,* 65–69.

Myers, J. E., Nell, V., Colvin, M., Rees, D., & Thompson, M. L. (1993). *Neuropsychological and neurological function in paint manufacturing workers with long-term exposure to organic solvents in two South African factories.* Unpublished data.

Nash, R. (1967). *Wilderness and the American mind.* New Haven, CT: Yale University Press.

Necka, E. (1992). Cognitive analysis of intelligence: The significance of working memory processes. *Personality and Individual Differences, 13,* 1031–1046.

Nell, V. (1985). Neuropsychological Screening Procedure. *Reports from the Psychology Department,* Whole No. 12. Pretoria: University of South Africa.

Nell, V. (1988). *Lost in a book: The psychology of reading for pleasure.* New Haven, CT: Yale University Press.

Nell, V. (1989). Health care and psychology in South Africa will be damaged by the introduction of middle level psychologists. Supplement to *Newsletter of the Psychological Association of South Africa,* 4/89, Pretoria.

Nell, V. (1990a). Rage dyscontrol: Part 1. Diagnostic and forensic issues. *South African Journal of Psychology, 20,* 235–242.

Nell, V. (1990b). Rage dyscontrol: Part 2. Approaches to treatment with special reference to traumatic brain injury. *South African Journal of Psychology, 20,* 243–249.

Nell, V. (1992). "Yes I said yes I will Yes": Review of Saayman, G. (Ed), *Modern South Africa in Search of a Soul: Jungian Perspectives on the Wilderness Within.* Boston: Sigo, 1990. *South African Journal of Psychology, 22,* 250–253.

Nell, V. (1994). Interpretation and misinterpretation of the South African Wechsler–Bellevue Adult Intelligence Scale: A history and a prospectus. *South African Journal of Psychology, 24,* 100–109.

Nell, V. (1996). Critical psychology and the problem of mental health. *Journal of Primary Prevention, 17,* 117–132.

Nell, V. (1997a). Science and politics meet at last: The South African insurance industry and neuropsychological test norms. *South African Journal of Psychology, 27,* 43–49.

Nell, V. (1997b). *Glasgow Coma Scale–Extended (GCS–E): Administration, scoring, and training manual* (Tech. Rep. No. 97/2). University of South Africa Health Psychology Unit, Johannesburg.

Nell, V. (1999). Standardising the WAIS–III and the WMS–III for South Africa: Legislative, psychometric, and policy issues. *South African Journal of Psychology, 29,* 128–137.

Nell, V., & Brown, D. S. O. (1990). *Epidemiology of traumatic brain injury in Johannesburg: Morbidity, mortality and etiology* (Tech. Rep. No. 90/1). Johannesburg: University of South Africa Health Psychology Unit.

Nell, V., & Brown, D. S. O. (1991). The epidemiology of traumatic brain injury in Johannesburg: II. Morbidity, mortality and etiology. *Social Science and Medicine, 33,* 289–296.

Nell, V., & Kirkby, L. (1997). *Psychometric research in Sub-Saharan Africa, 1900–1997: Sources and abstracts* (Tech. Rep. No. 97/2). Johannesburg: University of South Africa Health Psychology Unit.

Nell, V., Kruger, D. J., Taylor, T. R., Myers, J. E., & London, L. (1995). Bypassing culture: A performance process approach to the neuropsychological assessment of nonwestern subjects. Unpublished manuscript.

Nell, V., & Maboea, D. (1997). *Core battery of psychological and neuropsychological tests for persons with less than 12 years of education. Administration Procedure and Spoken Instructions* (Part 1: English; Part 2: South Sotho; Part 3: Zulu; Part 4: Afrikaans) (Tech. Rep. No. 97/4–8). Johannesburg: University of South Africa Health Psychology Unit.

Nell, V., Myers, J., Colvin, M., & Rees, D. (1993). Neuropsychological assessment of organic solvent effects in South Africa: Test selection, adaptation, scoring and validation issues. *Environmental Research, 63,* 301–318.

Nell, V., & Taylor, T. (1992). *Bypassing culture: Performance process probes of neuropsychological function in pesticide applicators* (Tech. Rep. No. 92/2). Johannesburg: University of South Africa Health Psychology Unit.

Nell, V., & Yates, D. W. (1998). An Extended Glasgow Coma Scale (GCS–E) with enhanced sensitivity to mild brain injury: 1. Rationale and field trials. 2. Administration, scoring, and training manual. Manuscript submitted for publication.

Ochse, R. (1990). *Before the gates of excellence: The determinants of creative genius.* Cambridge, England: Cambridge University Press.

Pang, D. (1985). Pathophysiologic correlates of neurobehavioural syndromes following closed head injury. In M. Y. Ylvisaker (Ed.), *Head injury rehabilitation: Children and adolescents* (pp. 3–70). London: Taylor & Francis.

Penn, C. (1988). The profiling of syntax and pragmatics in aphasia. *Clinical Linguistics and Phonetics, 2,* 179–207.

Piaget, J. (1926). *Language and thought of a child.* London: Routledge & Kegan Paul.

Pike, K. L. (1954). *Language in relation to a unified theory of the structure of human behavior.* Glendale, CA: Summer Institute of Linguistics.

Povlishock, J. T. (1996, January). *Role and pathobiology of axonal damage in traumatic brain injury.* Paper presented at the 1996 Advances in Acute Neurotrauma Conference, Philadelphia, PA.

Povlishock, J. T., & Jenkins, L. W. (1995). Are the pathobiological changes evoked by traumatic brain injury immediate and irreversible? *Brain Pathology, 5,* 415–426.

Prigatano, G. (1991). *Awareness of deficit after brain injury: Clinical and theoretical issues.* New York: Oxford University Press.

Psychological Corporation. (1997). *WAIS–III/WMS–III technical manual.* San Antonio: Author.

Reed, J. (1984). The contributions of Ward Halstead, Ralph Reitan and their associates. *International Journal of Neuroscience, 25*, 289–293.

Rees, L. M., Tombaugh, T. N., Gansler, D. A., & Moczynski, N. P. (1998). Five validation experiments of the test of memory malingering (TOMM). *Psychology Assessment, 10*, 10–20.

Reitan, R. M. (1969). *Manual for administration of Neuropsychological Test Batteries for adults and children.* Private publication.

Reitan, R. M. (1972). Verbal problem-solving as related to cerebral damage. *Perceptual and Motor skills, 34*, 515–524.

Reitan, R. M., & Davison, L. A. (Eds.). (1974). *Clinical neuropsychology: Current status and applications.* Washington, DC: Winston & Sons.

Reuning, H. (1988). Testing Bushmen in the central Kalahari. In S. H. Irvine & J. W. Berry (Eds.), *Human abilities in cultural context* (pp. 453–486). Cambridge, England: Cambridge University Press.

Reynolds, C. R. (Ed.). (1998). Common sense, clinicians, and actuarialism in the detection of malingering during head injury litigation. In *Detection of malingering during head injury litigation: Critical issues in neuropsychology* (pp. xii, 291). New York: Plenum.

Richardson, J. T. E. (1990). *Clinical and neuropsychological aspects of closed head injury.* London: Taylor & Francis.

Richter, E. D., Chuwers, P., Levy, Y., Gordon, M., Grauer, F., Marzouk, J., Levy, S., Barron, S., & Gruener, N. (1992). Health effects from exposure to organophosphate pesticides in workers and residents in Israel. *Israel Journal of Medical Science, 28*, 584–598.

Richter, L. (1990). *Wretched childhoods: The challenge to psychological theory and practice.* Inaugural address, University of South Africa.

Richter, L. M., & Griesel R. D. (1988). *Bayley Scales for infant development.* Pretoria: University of South Africa.

Richter, L. M., Griesel, R. D., & Wortley, M. E. (1988). The Draw-a-Man test: A 50 year perspective on drawings done by Black South African children. *South African Journal of Psychology, 19*, 1–5.

Richter-Strydom, L. M., & Griesel, R. D. (1984). *African infant precocity: A study of a group of South African infants from 2 to 15 months of age.* Pretoria: University of South Africa.

Rogan, J. M., & MacDonald, A. M. (1983). The effect of schooling on conversation skills. *Journal of Cross-Cultural Psychology, 14*, 309–322.

Rogers, R. (Ed.). (1997). Current status of clinical methods. In *Clinical assessment of malingering and deception* (2nd ed., pp. xii, 525). New York: Guilford.

Romer, C. J., von Holst, H., Gururaj, G., Kraus, J., Nell, V., Nygren, A., Servadei, F., Thurman, D., & Zitnay, G. (1995). *Prevention, critical care and rehabilitation of neurotrauma: Perspectives and Future strategies.* Stockholm: World Health Organization Collaborating Centers for Neurotrauma, Karolinska Institute.

Rosenstock, L., Keifer, M., Daniell, W. E., McConnell, R., & Claypoole, K. (1991). Chronic central nervous system effects of acute organophosphate pesticide intoxication. The Pesticide Health Effects Study Group. *Lancet, 27, 338*(8761), 223–227.

Ruff, R. M., & Crouch, J. A. (1991). Neuropsychological test instruments in clinical trials. In E. Mohr & P. Brouwers (Eds.), *Handbook of clinical trials: The neurobehavioral approach* (pp. 89–119). Amsterdam: Swets & Zeitlinger.

Russell, E. W., Neuringer, C., & Goldstein, G. (1970). *Assessment of brain damage: A neuropsycholigical key approach.* New York: Wiley.

Rutherford, W. H. (1989). Postconcussion symptoms: Relationship to acute neurological indices, individual differences, and circumstances of injury. In H. S. Levin, H. M. Eisenberg, & A. L. Benton (Eds.), *Mild head injury* (pp. 217–228). New York: Oxford University Press.

Sacks, O. (1989). *Seeing voices.* Berkeley: University of California Press.

Samuelson, L. (1994). All the news that that's fit to print and all the news that fits. *South African Journal of Science, 90*, p. 510 and p. 569.

Sarason, S. B. (1981). *Psychology misdirected.* New York: The Free Press.

Scarr, S. (1978). From evolution to Larry P., or what shall we do about IQ tests? *Intelligence, 2,* 325–342.

Scheerer, E., & Elliger, P. (1980). A bibliography of A. R. Luria's publications in the English, French, and German languages. *Psychological Research, 41,* 269–284.

Sesel, J. (1990). Mean times for two groups of South Africans for saying the alphabet, counting forward and backward, counting in threes, digit span, and controlled associate generation by letter of the alphabet and by category. Unpublished data, University of South Africa Health Psychology Unit.

Shallice, T. (1982). Specific impairments of planning. *Philosophical transactions of the Royal Society of London, 198,* 199–209.

Shochet, I. M. (1986). *Manifest and potential performance in advantaged and disadvantages students.* Unpublished doctoral dissertation, University of the Witwatersrand.

Smith, G. E., Ivnik, R. J., Malec, J. F., & Petersen, R. C. (1994). Mayo cognitive factor scales: Derivation of a short battery and norms for factor scores. *Neuropsychology, 8,* 194–202.

Snyderman, M., & Herrnstein, R. J. (1983). Intelligence tests and the Immigration Act of 1924. *American Psychologist, 38,* 986–995.

South African Clinical Neuropsychology Association (1995). Ethical principles for forensic psychologists. *SACNA Newsletter,* February, 2–3.

Spreen, O., & Strauss, E. (1991). *A compendium of neuropsychological tests: Administration, norms, and commentary.* New York: Oxford University Press.

Sternberg, R. J. (1984). Toward a triarchic theory of human intelligence. *The Behavioral and Brain Sciences, 7,* 269–315.

Sternberg, R. J. (1986). *The triarchic mind.* New York: Viking.

Sternberg, R. J. (1988). A triarchic view of intelligence in cross-cultural perspective. In S. H. Irvine & J. W. Berry (Eds.), *Human abilities in cultural context* (pp. 60–85). Cambridge, England: Cambridge University Press.

Sternberg, R. J. (1994). *Encyclopaedia of human intelligence.* (2 vols.). New York: Macmillan.

Sternberg, R. J., Conway, B. E., Ketron, J. L., & Bernstein, M. (1981). People's conceptions of intelligence. *Journal of Personality and Social Psychology, 41,* 37–55.

Sternberg, R. J., Wagner, R. K., Williams, W. M., & Horvath, J. A. (1995). Testing common sense. *American Psychologist, 50,* 912–927.

Tang, H. W., Liang, Y. X., Hu, X. H., et al. (1995). Alterations of monoamine metabolite and neurobehavioral function in leadworkers. *Biomedical and Environmental Sciences, 8,* 23–29.

Taylor, H. F. (1980). *The IQ game: A methodical inquiry into the heredity–environment controversy.* New Brunswick, NJ: Rutgers University Press.

Taylor, T. R. (1994). A review of three approaches to cognitive assessment and a proposed integrated approach based on a unifying theoretical framework. *South African Journal of Psychology, 24,* 184–193.

Teasdale, G., & Jennett, B. (1974). Assessment of coma and impaired consciousness. A practical scale. *Lancet, 2,* 81–84.

Tollman, S. G., & Msengana, N. B. (1990). Neuropsychological assessment: Problems in evaluating the higher mental functioning of Zulu-speaking people using traditional Western techniques. *South African Journal of Psychology, 20,* 20–24.

Triandis, H. C. (1995). *Individualism and collectivism.* Boulder, CO: Westview.

Trimble, J. E., Lonner, W. J., & Boucher, J. D. (1983). Stalking the wily emic: Alternatives to cross-cultural measurement. In S. H. Irvine & J. W. Berry (Eds.), *Human assessment and cultural factors* (pp. 259–274). New York: Plenum.

Tysvaer, A. T., Storli, O. V., & Bachen, N. I. (1989). Soccer injuries to the brain: A neurologic and electroencephalographic study of former players. *Acta Neurologica Scandinavica, 80,* 151–156.

UNESCO (1957). *World illiteracy at mid-century.* Paris: Author.

Verster, J. M. (1974). A study of intellectual structure in two groups of South African scientists. *Psychologia Africana, 15,* 169–190.

Verster, J. M. (1983). The structure, organisation, and correlates of cognitive speed and accuracy: A cross-cultural study using computerised tests. In S. H. Irvine & J. W. Berry (Eds.), *Human assessment and cultural factors* (pp. 175–292). New York: Plenum.

Verster, J. M. (1986). *Cognitive competence in Africa and models of information processing: A research prospectus* (Report No. Pers-411). Pretoria: Human Sciences Research Council.

Verster, J. M. (1991). Simon Biesheuvel. *South African Journal of Psychology, 21,* 267–270.

Verster, J. M., & Muller, M. W. (1985). *Further investigation of the effects of multiple exposure to the classification test battery: Hypotheses and proposed research design* (Confidential Report No. c/Pers 231). Johannesburg: National Institute for Personnel Research.

Verster, J. M., & Prinsloo, R. J. (1988). The diminishing test performance gap between English speakers and Afrikaans speakers in South Africa. In S. H. Irvine & J. W. Berry (Eds.), *Human abilities in cultural context* (pp. 534–559). Cambridge, England: Cambridge University Press.

Verster, M. A. (1976). *The effect of mining experience and multiple test exposure on test performance of Black mine workers.* Pretoria: University of South Africa.

Vygotsky, L. S. (1988). *Mind in society: The development of higher psychological processes.* Cambridge, MA: Harvard University Press.

Walsh, K. W. (1985). *Understanding brain damage: A primer of neuropsychological evaluation.* Edinburgh: Churchill Livingstone.

Walsh, K. W. (1987). *Neuropsychology: A clinical approach.* Edinburgh, England: Churchill Livingstone.

Weber, M. (1904/1965). *The Protestant Ethic and the spirit of capitalism.* London: Unwin.

Wechsler, D. (1945). A standardised memory scale for clinical use. *Journal of Psychology, 19,* 87–95.

Wechsler, D. (1958). *The measurement and appraisal of adult intelligence* (4th ed.). Baltimore: Williams & Wilkins.

Wechsler, D. (1981). *Administration and scoring manual for the Wechsler Adult Intelligence Scale–Revised (WAIS–R).* San Antonio, TX: The Psychological Corporation.

Wechsler, D. (1997). *Administration and scoring manual for the Wechsler Adult Intelligence Scale* (3rd ed.). San Antonio, TX: The Psychological Corporation.

Werner, D., & Sanders, D. (1996). *Questioning the solution: The Politics of Primary Health Care and Child Survival.* Palo Alto, CA: HealthWrights.

Wilson, B. A., Baddeley, A., Shiel, A., & Patton, G. (1992). How does post-traumatic amnesia differ from the amnesic syndrome and chronic memory impairment? *Neuropsychological Rehabilitation, 2,* 231–243.

Wober, M. (1969). Meaning and stability of Raven's Matrices Tests among Africans. *International Journal of Psychology, 4,* 229–235.

Woody, S. (1988). Episodic dyscontrol syndrome and head injury. *Journal of Neurosciences and Nursing, 20,* 180–184.

World Health Organization (1986). *Operational guide for the WHO Neurobehavioural Core Test Battery.* Office of Occupational Health. Geneva: Author.

Yeudall, L. T., Fromm, D., Reddon, J. R., & Stefanyk, W. O. (1986). Normative data stratified by age and sex for 12 neuropsychological tests. *Journal of Clinical Psychology, 42,* 919–947.

Zindi, F. (1994). Differences in psychometric performance: Zimbabwe and England. *The Psychologist, 7,* 549–552.

# Author Index

D



E



F



G



H

## S



## T



## U



## V



## W



## Y



## Z

# Subject Index

## A

## D

## F



## G



## H

## I



## J

## K



## L



## M



## N

## T

## U



## V